Advanced Controls for Wind Driven Doubly Fed Induction Generators

Advanced Controls for Wind Driven Doubly Fed Induction Generators discusses the most advanced control algorithms used for enhancing the dynamics of a doubly fed induction generator (DFIG) operating at fixed and variable speeds, and which are used for different utilization purposes (standalone and grid connection). Extensive generator performance analysis has been introduced using various control topologies.

Features

- Presents modeling of wind energy conversion systems (WECS), including a wind turbine as a prime mover, a DFIG as a generation unit for electrical energy, and a three-phase induction motor as an isolated load
- Explores a detailed description for the presented control algorithms in order to visualize the base principle of each method
- Introduces a comprehensive performance analysis for the DFIG using the formulated predictive voltage control scheme and other control techniques under different operating conditions
- Examines the formulation of new control approaches which overcome the shortages present in previous DFIG control schemes
- Presents a detailed comparison between different control topologies for the DFIG to outline the most effective procedure in terms of dynamic response, structure simplicity, ripples, total harmonic distortion, and computational burdens

The book is written for researchers and academics working on advanced control systems and those interested in areas such as machine drives, renewable energy systems, 'adaptive control,' modeling of WECS, and optimization theory.

Advanced Controls for Wind Driven Doubly Fed Induction Generators

Mahmoud K. Abdelhamid, Mahmoud A. Mossa, Ahmed A. Hassan

CRC Press
Taylor & Francis Group
Boca Raton London New York

CRC Press is an imprint of the
Taylor & Francis Group, an **informa** business

MATLAB® and Simulink® are trademarks of The MathWorks, Inc. and are used with permission. The MathWorks does not warrant the accuracy of the text or exercises in this book. This book's use or discussion of MATLAB® and Simulink® software or related products does not constitute endorsement or sponsorship by The MathWorks of a particular pedagogical approach or particular use of the MATLAB® and Simulink® software.

Designed cover image: © Shutterstock Images

First edition published 2024
by CRC Press
6000 Broken Sound Parkway NW, Suite 300, Boca Raton, FL 33487-2742

and by CRC Press
4 Park Square, Milton Park, Abingdon, Oxon, OX14 4RN

CRC Press is an imprint of Taylor & Francis Group, LLC

© 2024 Mahmoud K. Abdelhamid, Mahmoud A. Mossa, Ahmed A. Hassan

ISBN: 9781032572901 (hbk)
ISBN: 9781032576855 (pbk)
ISBN: 9781003440529 (ebk)

DOI: 10.1201/9781003440529

Typeset in Times
by codeMantra

In the Name of Allah, the Most Beneficent, the Most Merciful

*Dedicated to the soul of my dear parents, sisters, wife,
and my son Mohamed, who represent life for me.*

Mahmoud K. Abdelhamid, 2023

*Dedicated to my father, mother, wife and
my kids who are the light of my eyes.*

Mahmoud A. Mossa, 2023

Dedicated to my family especially my parents, wife, sons, and girls.

Ahmed A. Hassan, 2023

Contents

Preface

As a cost-effective, environmentally friendly, and secure alternative to nuclear and fossil fuel power generation, wind electrical power systems are currently receiving a lot of attention. High-power wind applications frequently employ a unique kind of generator known as doubly fed induction generator (DFIG). Due to its simple controllability, improved efficiency, and increased power quality, it is being employed more and more in wind turbine applications. These applications involve the grid connection and standalone operation as well. Accordingly, searching for the most effective control to be used with the DFIG became a challenge for the researchers in this field of study. The majority of the adopted control approaches with the DFIG are formulated based upon utilizing the pulse width modulation (PWM) with linear regulators. All of these facts resulted in making the overall control scheme more complex. Recent studies have shown that the DFIG can be controlled using an advanced control theory which can be adopted without modulators or linear controllers, and all of these requirements are fulfilled by the predictive control. This is adaptable enough to incorporate adjustments and extensions of control horizons according to the required applications, the capacity to take into account a multi-objective scenario inside the model, simple system constraints handling, straightforward integration of non-linearities inside the model, and straightforward digital implementation.

Accordingly, the study presented in this book is investigating in detail the application of predictive control as an advanced approach with its different forms to manage the dynamic performance of a DFIG for standalone operation purposes. The performance is evaluated for the direct-driven and wind-driven operation as well. All generation system components are initially modeled. First, the detailed mathematical model of the DFIG is introduced in the general rotating frame to give extra freedom when applying the control. After that, the wind turbine system and its power management unit, which adopts the maximum power point tracking (MPPT) principle to optimally exploit the wind energy, are modeled and described. Following that, the mathematical model of the isolated load to be fed by the DFIG under standalone operation is described. The load under study is a three-phase induction motor (IM) as an emulator to a realistic loading condition.

To perform a detailed performance comparison, the study applied, first, the traditional vector control, specifically, the stator voltage-oriented control (SVOC) type with both direct and wind-driven DFIG. The control design is described in detail, and the results are extracted. After that, the study formulated different forms of predictive control to be used with the DFIG; some of them were recently used, and the others are newly designed. The recently used predictive control topologies are the model predictive direct torque control (MPDTC) and model predictive current control (MPCC); meanwhile the new designed predictive scheme is given the title of 'predictive voltage control (PVC).' Alternatively, for controlling the operation of the isolated load (IM), the MPDTC principle is used. The design of all predictive controllers is articulated mainly on the type of used cost function as illustrated and described in detail within the book.

As an additional verification for the performances of all adopted control schemes with the DFIG used in standalone operation cases and to assure the superiority of the designed PVC scheme over the other controllers, the dynamics of the DFIG are also evaluated under grid connection state. All test scenarios are carried out using the MATLAB/Simulink software.

Extensive performance evaluation tests are performed, which in conclusion report that the performance of the newly designed PVC control exhibits the most effective one compared with the other controllers. Specifically, the designed PVC has a faster response time and simpler structure compared with the classic SVOC; meanwhile it has faster dynamics, lower torque and flux ripples, lower current harmonics, and simpler structure than the MPDTC and MPCC schemes. These facts are confirmed under both standalone and grid connection states while driving the generator shaft through two ways: directly and using wind turbine as well.

MATLAB® is a registered trademark of The MathWorks, Inc. For product information, please contact:

The MathWorks, Inc.
3 Apple Hill Drive
Natick, MA 01760-2098 USA
Tel: 508-647-7000
Fax: 508-647-7001
E-mail: info@mathworks.com
Web: www.mathworks.com

Authors' Biographies

Mahmoud K. Abdelhamid received a BSc degree in Electrical Engineering in 2019 from the Electrical Engineering Department, Faculty of Engineering, Minia University, Egypt. Then, he was nominated as a Teaching Assistant in the same department. Further, he also received an MSc degree in Electrical Engineering from the Electrical Engineering Department, Minia University in 2023. Recently, he has been working as an Assistant Lecturer in the same department. His research interests focus on renewable energy systems, electric machine drives, and control systems.

Mahmoud A. Mossa received bachelor's and master's degrees in electrical engineering from the Faculty of Engineering, Minia University, Egypt, in 2008 and 2013, respectively, and a PhD degree in electrical engineering, in April 2018. Since January 2010, he has been working as an Assistant Lecturer at the Electrical Engineering Department, Minia University. In November 2014, he joined the Electric Drives Laboratory (EDLAB), University of Padova, Italy, for his PhD research activities. Since May 2018, he has been working as an Assistant Professor at the Electrical Engineering Department, Minia University. He occupied a Postdoctoral Fellow position at the Department of Industrial Engineering, University of Padova for six months during the academic year 2021/2022. He is an Associate Editor in *International Journal of Robotics and Control Systems*. His research interests include renewable energy systems, power management, control systems, electric machine drives, power electronics, optimization, and load frequency control.

Ahmed A. Hassan received a BSc and an MSc degree from Assiut University, Egypt, and a PhD degree from Minia University, Egypt, all in electrical engineering, in 1976, 1982, and 1988, respectively. Since 2001, he has been a Professor with the Electrical Engineering Department, Minia University. He was also the Dean of the faculty of Computer Science and Information in the same university. His current interests include the application of advanced control techniques to electrical machine drives.

Acronyms

CCS PC Continuous control set predictive control
DBPC Deadbeat-Based Predictive Control
DFOC Direct field-oriented control
DPC Direct power control
DTC Direct torque control
DFIG Doubly fed induction generator
FOC Field orientation control
FCS PC Finite control set predictive control
IFOC Indirect field-oriented control
IG Induction generator
IM Induction motor
MPPT Maximum power point tracking
MPC Model predictive control
MPCC Model predictive current control
MPDTC Model predictive direct torque control
PMSG Permanent magnet synchronous generator
PLL Phase-locked loop
PC Predictive control
PVC Predictive voltage control
PI Proportional integrator
PWM Pulse width modulation
SEIG Self-excited induction generator
SCIG Squirrel cage induction generator
SCIM Squirrel cage induction motor
SFOC Stator field-oriented control
SVOC Stator voltage-oriented control
SG Synchronous generator
SRF Synchronous reference frame
TSR Tip speed ratio
THD Total harmonic distortion
VOC Vector orientation control
VSI Voltage source inverter
WECS Wind energy conversion system
WPSs Wind power systems
WT Wind turbine
WRSG Wound rotor synchronous generator

NOMENCLATURE

ρ Air density, kg/m^3
T_{me} & T_g Applied mechanical torque to the DFIG's shaft, Nm
β Blade pitch angle, degree

R	Blade radius of wind turbine, m
D	Damping factor
T_d	Developed torque by DFIG, Nm
J_g	DFIG's inertia, kg.m²
$\omega_{m,k}$	Electrical angular speed of IM, rad/s
ω_{me}	Electrical rotor angular speed, rad/s
F	Friction constant
G	Gear ratio
J_{IM}	IM's inertia, kg.m²
σ	Leakage factor
$T_{l,k}$	Load torque applied to IM's shaft, Nm
$\omega_{mech_{IM}}$	Mechanical angular speed of IM, rad/s
ω_{mech} & ω_g	Mechanical rotor angular speed of DFIG, rad/s
L_m	Mutual inductance of DFIG, H
L_{mm}	Mutual inductance of IM, H
ω_n	Natural frequency of the system, rad/s
λ_{opt}	Optimal value of TSR
p	Pole pairs of DFIG
p_{IM}	Pole pairs of IM
C_p	Power coefficient of wind turbine
ω_g^*	Reference value of generator speed, rad/s
ω_t^*	Reference value of turbine speed, rad/s
$\bar{i}_r^{sv}$	Rotor current vector, A
$\bar{\Psi}_r^{sv}$	Rotor flux vector, Vs
L_r	Rotor inductance of DFIG, H
L_{lr}	Rotor leakage inductance of DFIG, H
$L_{lr,m}$	Rotor leakage inductance of IM, H
L_{rm}	Rotor leakage inductance of IM, H
θ_{me}	Rotor position, degree
R_r	Rotor resistance of DFIG, Ω
R_{rm}	Rotor resistance of IM, Ω
$\bar{u}_r^{sv}$	Rotor voltage vector, V
T_s	Sampling time, s
ω_{slip}	Slip angular speed, rad/s
P_s	Stator active power, Watt
$\bar{i}_s^{sv}$	Stator current vector, A
$\bar{\Psi}_s^{sv}$	Stator flux vector, Vs
L_s	Stator inductance of DFIG, H
L_{sm}	Stator inductance of IM, H
L_{ls}	Stator leakage inductance of DFIG, H
$L_{ls,m}$	Stator leakage inductance of IM, H
Q_s	Stator reactive power, Var
R_s	Stator resistance of DFIG, Ω
R_{sm}	Stator resistance of IM, Ω
L_t	Stator transient inductance of DFIG, H

$\theta_{\bar{u}_s}$	Stator voltage angle, degree
$\bar{u}_s^{sv}$	Stator voltage vector, V
A	Swept area of wind, m²
θ_s	Synchronous angle, degree
$\omega_{\bar{u}_s}$	Synchronous angular speed, rad/s
λ	Tip speed ratio
T_e	Torque developed by IM, Nm
J_t	Turbine inertia, kg.m²
P_t	Turbine power, Watt
ω_t	Turbine speed, rad/s
T_t	Turbine torque, Nm
ω_f	Weighting factor
P_ω	Wind power, Watt
V_ω	Wind speed, m/s

Introduction

0.1 HISTORICAL BACKGROUND

To maximize the power extraction from renewable energy sources including wind, geothermal, solar, and wave energies, developers must consider novel control approaches due to the rising need for clean energy sources in our daily life [1–5]. Electric generators used in wind energy conversion systems come in a variety of topologies, including SGs, SEIGs, and DFIGs [6–13]. Two categories of wind turbines are used in WECS: fixed-speed wind turbines and variable-speed wind turbines. In fixed-speed wind turbines, a synchronous generator can be used where the speed control can be performed with the aid of gear ratio of the gear box in between the turbine and the generator's shaft; but this type has a substantial defect: the mechanical losses are very high which leads to reducing the efficiency of the energy conversion process. However, variable-speed wind turbines use DFIG with back-to-back converter in between stator and rotor windings, which eliminates the need to gear ratio control. Due to its various benefits, including the ability to provide output voltages with constant amplitude and frequency regardless of changes in wind speed, the DFIG has been employed with wind turbines frequently recently and is still in use today [14–16]. The flexibility of the control, which allows it to be operated from either the stator side or the rotor side, is another important aspect of the DFIG [17]. The rating of the power electronic components used in the inverter is reduced as a result of controlling the DFIG from the rotor side because they only need to process the slip power, which reduces converter losses and, consequently, the overall cost [18, 19].

Researchers have worked hard to create different control strategies and then test them with the DFIG due to the DFIG's many benefits as a generation unit. While DTC technique is utilized in [20, 21], VOC technique is used in [22, 23]. Various versions of the VOC approach have been used depending on the variable that was chosen to be aligned with the direct (d) axis of the rotating synchronous frame. For example, in [24], the stator field was selected to rotate at the same rate as the synchronous frame, and in [23], the stator voltage was selected to be in line with the direct axis of the synchronous frame. The rotor field in [25, 26] was chosen to be aligned with the direct axis of the synchronous frame. The DFIG experiences a smooth dynamic response under VOC as a result of the VOC schemes' success in enhancing torque response and minimizing ripples. The VOC systems still have a few shortages, such as the fact that they rely on machine parameters, which are used for estimating the control variables that might be affected by ambient variables like temperature fluctuation. In addition, they need to coordinate transformations and PI regulators, which complicates the system and delays the dynamic response. They also need to use a modulation stage.

DOI: 10.1201/9781003440529-1

Hysteresis torque and rotor flux comparators were used in place of the PI regulators to create the DTC of DFIGs in an effort to reduce the complexity of the VOC structure [21]. Although DTC has the benefit of quick responses, the ripples problem still poses the biggest obstacle [27]. PC was recently discovered as a result of academics' attempts to provide a new control strategy that eliminates the shortcomings of VOC and DTC [28–30]. It substitutes a cost function for the PI regulators, current control loops, and hysteresis comparators seen in VOC and DTC. This has made it possible to create a quick dynamic response, produce a superior steady-state performance, and reduce complexity and ripples. The chosen cost function has the benefit of flexibility and the ability to handle numerous control objectives at once. It comprises two components that assess the absolute inaccuracy of the variables that must be regulated. MPC is used in a wide range of applications, such as [31] to improve the performance of the power system and [32] to work in tandem with a microgrid system.

MPDTC, which adopted FCS PC, has been utilized to address various DTC flaws including increasing ripples. The goal of MPDTC is to reduce the difference in torque and rotor flux between the reference and actual values. Unfortunately, MPDTC has some flaws. At first, the cost function needs a ω_f to attain the desired equilibrium between torque and rotor flux. Furthermore, any improper selection of ω_f exposes the torque and rotor flux to an increase in ripples. Another flaw in MPDTC is that because the cost function's components must be calculated, any discrepancy or adjustment to the model's parameters will affect the estimated variables. Additionally, MPDTC's computational weight is viewed as a problem since it necessitates the estimate and prediction of rotor flux and torque [33–35].

Multiple vectors were implemented within the same sample interval in order to address MPDTC's shortcomings, particularly the apparent ripples [36, 37]. Although the computation time rose once more, this step significantly reduced the flux and torque fluctuations near their references. To obtain the best balance between the flux and torque variations, some researchers utilized an online adaption of the weighting factor [38]. The online weighting update produced a good performance, but it increased the computational load on the system, making it unsuitable for all situations where the execution capacity of the microprocessors is constrained. To obtain smooth flux and torque variations, several investigations looked into the insertion of adaptive flux estimators [39, 40]. The adaptive estimators produced accurate findings, but the complexity of the system suffered as a result.

MPCC, a different PC topology that attempts to fix some of MPDTC's flaws, exhibits less torque and flux ripples overall than MPDTC. Additionally, the cost function of this controller is simple and does not require a weighting factor scale. Regarding the disposal of the PI regulators, which are frequently used by the VOC technique, MPCC and MPDTC are in agreement. Unfortunately, MPCC's response time is longer than MPDTC's response time, and it still has a heavy computational burden. Additionally, the variables in its cost function must be evaluated using a machine model, which is easily impacted by operating conditions [41–44].

0.2 BOOK MOTIVATIONS

As the availability of the non-renewable energy resources is decreasing day by day, there is a severe need for finding alternative renewable energy resources and consequently searching for the most appropriate control approach which can extract the maximum available power from these resources, one of these resources is the wind energy which became a common resource in all over the world and also here in Egypt. Accordingly, in our study, we started to study a complete generation system based on wind energy as a source and analyze this system from the control point of view. As the DFIG is considered one of the most commonly used generators with wind turbines, which is known as WECS, so we studied the dynamic performance of this generator in our study and evaluated its performance under different operating conditions and using various control approaches. Usually, finding the optimal control strategy to be used with non-linear system require performing a comprehensive and detailed dynamic performance analysis; to achieve this requirement in our study, we adopted different categories of control theories, starting from classic controllers passing until reaching to the most recent controllers which is the predictive controller. Additionally, the PC itself uses different mechanisms according to the way by which each controller works, which depends mainly on the cost function "convergence condition." Accordingly, in our study, we intended to design a new predictive control scheme with a unique cost function "convergence condition" which can achieve a better performance compared with other controllers.

0.3 BOOK OBJECTIVES

Given the previously mentioned succinct discussion, the book's objectives can be listed as follows:

- Modeling of the wind energy conversion system includes a DFIG as a generation unit for electrical energy, a wind turbine as a prime mover, and a three-phase induction motor as an isolated load.
- Enhancing and controlling the dynamic performance of a wind-driven DFIG during variable wind speeds using advanced control techniques.
- Introducing a detailed description for the adopted control algorithms in order to visualize the base principle of each method, showing when it works properly and when it fails.
- Formulating an effective PVC approach which overcomes the shortages present in previous DFIG control schemes.
- Presenting a comprehensive performance analysis for the DFIG using the formulated PVC scheme and other control techniques under different operating conditions.
- Performing a fair and detailed comparison between different control topologies for the DFIG to outline the most effective procedure in terms of dynamic response, structure simplicity, ripples, THD, and computational burdens.

0.4 BOOK OUTLINES

The present book is organized in five chapters, as follows:

Chapter 0: "Introduction" outlines the historical context of using induction machine drives and the various control methods employed for them. Additionally, it displays the book's motives, goals, and outlines.

Chapter 1: "Literature Review" gives a quick overview of the many kinds of wind generation systems and the different kinds of generators employed in each system. A literature evaluation of various controls for a DFIG that is powered by the wind is offered. Moreover, it presents the model predictive control strategy and explains its operating principle in a detailed manner.

Chapter 2: "Modeling of the Wind Energy Conversion System Components" introduces a complete modeling for the wind energy conversion system components, including a DFIG as a generation unit for electrical energy, a wind turbine as a prime mover, and a three-phase induction motor as an isolated load.

Chapter 3: "Analyzing and Enhancing the Performance of a Standalone Variable-Speed DFIG" introduces the detailed mathematical model of the DFIG, then, the wind turbine system and its power management unit are modeled and described. After that, the mathematical model of the adopted isolated load which is supplied by the generator is presented. Furthermore, the dynamics of the DFIG under various control approaches is analyzed, presenting the operating principle of each technique, following that, the proposed controller is formulated and introduced in a detailed and systematic manner. Finally, a comprehensive dynamic performance comparison is carried out among the adopted controllers to clarify which controller is the most suitable to be used with the DFIG.

Chapter 4: "Dynamic Performance Analysis of an Alternative Operating Regime of DFIG (Grid Connected Case)" tests the performance of a grid-connected DFIG under variable operating conditions with the proposed predictive controller. Also, the performance of the DFIG is evaluated using other control techniques to verify the validation of the designed control approach. Lastly, a comprehensive comparison is implemented between the controllers which are adopted; through which the merits and flaws of each algorithm are listed.

Chapter 5: "Conclusions and Recommendations for Future Work" introduces the book's primary contributions and results, as well as suggestions for future research.

1 Literature Review

1.1 INTRODUCTION

The most common form of energy in our homes, businesses, and workplaces is electrical energy. Over the past three decades, there has been a noticeable rise in electricity demand due to population and industrial growth. Natural resources like coal, gas, and petroleum that have long powered our power plants, industries, and automobiles are quickly running out. This critical issue has prompted countries all over the world to consider alternate energy sources that make use of non-depletable natural resources. The constant progress in technology over the previous 20 years has helped wind farms remarkably. The electrical machines, control capabilities, and power converters are among the components where there has been a significant amount of progress. Simple induction machines with soft starts are a thing of the past. We are now able to limit power output and regulate voltage and speed, as well as the machine's real and reactive power [45]. Around the world, there is a lot of research being done in this area, and technology that has a lot of potential is being created. It necessitates a working knowledge of machines, power systems, control schemes, and applications of power electronic converters. Wind turbines (WTs) can run at a constant speed or a variable speed, unlike a traditional power plant that uses synchronous generators which must be driven by constant speeds. Use of a variable-speed WT is justified for a number of reasons:

  i. WTs which turn with variable speeds produce more energy than those turn with fixed speeds.
  ii. Variable-speed operation can reduce mechanical strains and provide straightforward pitch control.
  iii. WTs with variable speeds provide substantial noise reduction and both active and reactive power are controllable.
  iv. WTs with variable speeds exhibit less output power variability [46, 47].

In addition to supplying the rising demand for electricity from consumers, the use of renewable energy sources for the production of electric power is becoming more and more important in the fight against climate change and environmental pollution. Grid integration of wind energy electric conversion systems is among the many renewable energy technologies that are being installed in great numbers because they convert energy cleanly and cheaply. Recent advancements in WT and power electronics technology have also made it easier to quickly integrate WECSs into the grid [48–50].

There are two types of WTs according to the axis of rotation: horizontal axis WT and vertical axis WT. For horizontal axis WTs, its blades spin around a horizontal axis parallel to the wind speed, also, it has higher speeds because the blades are far from ground. For vertical axis WTs, its blades spin around a vertical axis, and it has lower speeds because the blades are close to ground.

DOI: 10.1201/9781003440529-2

Three different converter types are frequently utilized with WECS. These converters are matrix, multi-level, and two-level converters. The matrix converter may convert the generator's variable AC frequency directly into the grid's fixed AC frequency. It is an AC-AC converter as a result. With this design, the converter only needs to conduct control, therefore there are no bulky energy storage or DC-link requirements. Compared with the traditional two-level topology, multilevel converters are AC-DC-AC converters that are well suited for medium- and high-power applications because they can meet the rising demand for power ratings and power quality associated with reduced harmonic distortion, lower electromagnetic interference, and higher efficiencies. Back-to-back PWM converter is the term often used to describe two-level power converters, and this is the type which we adopted in our book.

The main generator types for variable-speed WPSs are covered in the following sections.

1.2   WIND-DRIVEN SG

SGs typically consist of a rotor that serves as the magnetic field source and a stator, which houses the three-phase windings used to power the external load. The rotor may be a permanent magnet type or a wound rotor type which can be supplied by direct current through the field winding.

1.2.1   WIND-DRIVEN WRSG

A four-quadrant power converter with two back-to-back sinusoidal PWM connects the stator winding to the network. The grid side converter (GSC) controls the actual and reactive power that the WPS delivers to the utility, while the machine side converter (MSC) controls the electromagnetic torque. Some benefits of the wound rotor synchronous generator include the following:

- Due to the fact that it uses the entire stator current to generate electromagnetic torque, this machine often has a high efficiency [49].
- The fundamental advantage of using a wound rotor synchronous generator with a salient pole is that it enables direct control of the machine's power factor, allowing the stator current to be reduced under any operating conditions.

However, compared with a PMSG, the rotor's winding circuit may be a disadvantage [51].

1.2.2   WIND-DRIVEN PMSG

A PMSG had been employed in a variety of configuration designs. One of them had a boost converter connected to a three-phase rectifier that was connected to a PMSG. The boost converter in this instance regulates the electromagnet torque. The drawback of using a diode rectifier in this configuration is the increase in PMSG's

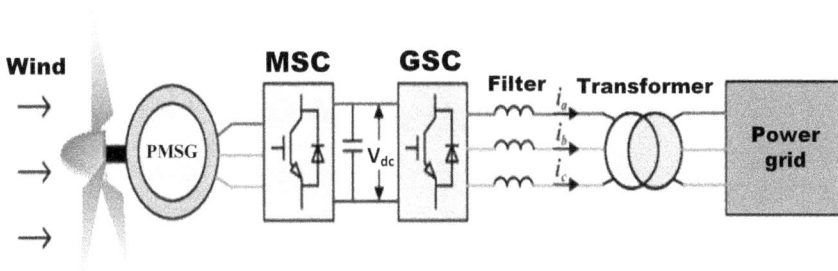

FIGURE 1.1 Configuration of WECS based on PMSG.

current amplitude and distortion [52]. This configuration has therefore been taken into consideration for small-scale WPSs (less than 50 kW).

Another PMSG system connects a PWM inverter (GSC) to the network while positioning a PWM rectifier (MSC) among the generator and the DC link, as shown in Figure 1.1. The benefit of using VOC with this system is that it enables the generator to run close to its ideal working point, minimizing losses in the generator and power electronic circuit. However, the effectiveness depends on having a solid understanding of the generator parameter, which changes depending on frequency and temperature. The main disadvantages of using PMSG are the permanent magnets' cost, which drives up the cost of the machine, permanent magnet material demagnetization, and the inability to regulate the machine's power factor [53, 54].

1.3 WIND-DRIVEN IG

IGs have historically been the most popular type of AC generators utilized in WTs. The two types of IGs most frequently used in WTs are SCIG and DFIG.

1.3.1 WIND-DRIVEN SCIG

In standalone power systems that use non-conventional energy sources such as wind energy and hydropower, three-phase SCIGs are frequently used. This is because of these generators' benefits over traditional SGs. Less expensive per unit, absence of a separate excitation d.c. source, robustness, ease of maintenance, and brushless rotor construction are the key benefits. The machine can be used as an SEIG if the rotor winding is driven at a suitable speed, and an appropriate three-phase capacitor of suitable value is connected across its stator terminals. Also, SCIG can be connected to the grid as illustrated in Figure 1.2, in which two back-to-back converters coupled with a DC connection, connect the stator winding of this generation system to the grid.

The machine side converter's management system manages the torque and provides reactive power to keep the machine magnetization [55]. The grid side converter controls the real and reactive power sent from the system to the utility and controls

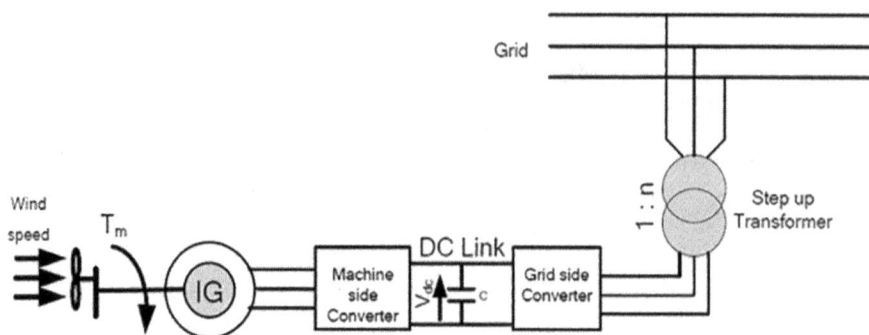

FIGURE 1.2   Wind-driven SCIG.

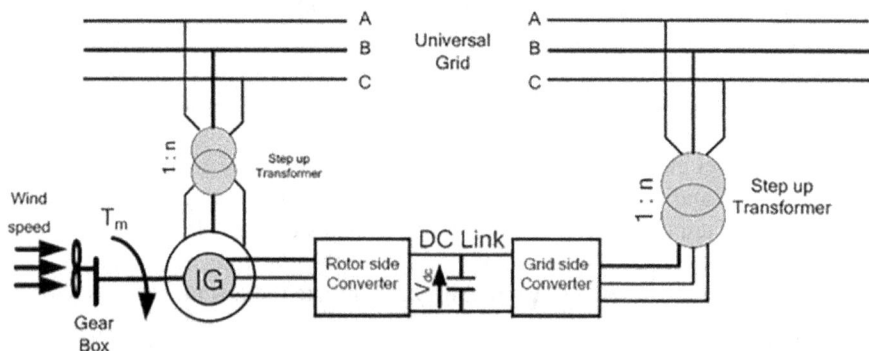

FIGURE 1.3   Wind-driven DFIG.

the DC connection, although there are certain disadvantages to using squirrel cage induction generators as follows:

- To meet the machine's need for magnetizing, the stator side converter must be larger by 30%–50% of its rated power [55].
- It needs a complex control system, and its effectiveness depends on an in-depth understanding of the parameters of the generator such as frequency, temperature, and magnetic saturation [55].

1.3.2   WIND-DRIVEN DFIG

The WPS in Figure 1.3 consists of a DFIG, whose stator winding is connected to the grid directly and whose rotor winding is connected to the grid via two back-to-back converters. The rotor side converter's controller typically controls the electromagnetic torque and provides some reactive power to keep the machine's magnetization in place. However, the DC link voltage is regulated by the stator side converter's controller [56]. Wind-driven DFIGs may be exposed to faults while connecting to the grid-like over-voltage fault on the DC link capacitor, which requires a protection technique to protect the system during the fault period. In [57], there are two different kinds of crowbar protections used: one in the DC link and the other in the rotor winding. The rotor winding crowbar links in series with the rotor winding and rotor MSC during the fault situation to reduce the MSC current and disperse the rotor energy. The over-current cannot be greatly reduced by the general parallel rotor side crowbar (PRSC). Under severe faults, DFIG shouldn't be kept linked to the utility grids in order to protect the semiconductor switches of RSC. Only when the DC capacitor voltage rises over a certain point does the DC link capacitor crowbar (DCCC) turn on.

These benefits of this DFIG over SGs include the following:

- Because inverters normally provide 25% of the system's power, their costs are reduced. This is so that the converters just have to manage the rotor's slip power.
- Because inverter harmonics make up a lesser portion of the overall system harmonics and inverter filters are rated for 0.25 p.u. of system power, they are less expensive.
- The machine is more stable and robust in face of external disturbance [56].

1.4   FIELD ORIENTATION CONTROL (FOC) OF DFIG

The control system of a variable-speed WT with DFIG aims to control the reactive power exchanged between the generator and the grid as well as the active power drawn from the WT in order to track the WT's optimal operation point or to limit the power in the case of high wind speeds. Each WT system consists of three subsystems: mechanical, electrical, and aerodynamical. The electrical dynamics are often significantly faster than the mechanical dynamics. Due to the existence of the power electronics in a variable speed WT, this discrepancy in time constants becomes even more. A control system that is more sophisticated is also required for such a complicated electrical system. Two decoupled control channels are presented in the DFIG control system, one for the stator side converter and the other for the rotor side converter. Each of these control channels generates a PWM factor for the corresponding converter. Nowadays, DFIGs constitute the basis for many variable speed WTs, which are connected to the grid or an isolated load via back-to-back converters. The main benefit of these facilities is that the inverters' power rate is approximately 25% of the generator rated power. This feature allows for the regulation of electrical power generation within this range, which has been shown to be a reasonable trade-off between cost-effective operation and operation efficiency.

VOC algorithms often serve as the foundation for most common power control systems which are utilized with DFIG-WTs. In this method of control, the injection of active and reactive powers is controlled separately using the d-q components of the rotor currents which obtained using Park's transformations [58, 59]. The SRF transforms enable the separate control of P and Q. Due to how simple it is to adjust the controller parameters, this functionality has begun to be used in many DFIG-WTs. The main idea behind this control is to align one of the machine's fluxes with an orthogonal d-q system created from the three-phase variables in an AC machine. As a result, the machine's flux and electromagnetic torque can be controlled independently. Induction machines employ both DFOC and IFOC as methods of field

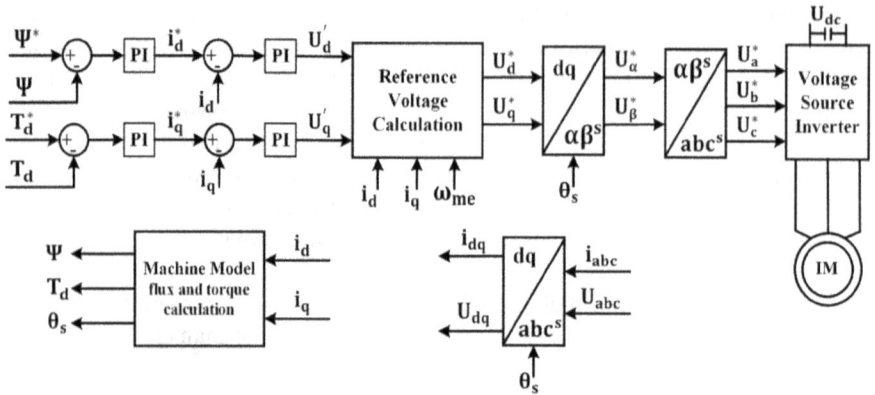

FIGURE 1.4   SFOC for induction machine.

orientation. The IFOC, which is frequently used in motor drives and generator applications, can work in four quadrants down to standstill. Usually, the flux of the rotor is aligned with the orthogonal SRF. However, the machine's parameters have a significant impact on this control. The DFOC with SFOC is less susceptible to machine parameters and does not require knowledge of the rotor speed. However, it exhibits poor performance at slow speeds close to a standstill.

Figure 1.4 depicts a general control structure for SFOC for induction machines in SRF. The d-axis regulates the machine's flux or reactive power, while the q-axis manages the machine's electromagnetic torque or active power. Based on the machine equations and the currents, the actual flux, torque, and flux angle are calculated. For the DFIG systems, a similar control mechanism is employed. The active and reactive powers on the stator side of the machine are typically controlled by the outer control loops.

Following is a comparison of the two fundamental field orientation schemes DFOC and IFOC:

The DFOC requires flux acquisition (position and magnitude), which is mostly accomplished through computational techniques using machine terminal values. IFOC avoids the direct flux acquisition by combining the shaft speed with an approximated and controlled slip frequency to determine the rotor flux position [60].

1.4.1   DFOC OF A DFIG DRIVEN BY A WT

In DFOC, the position of the rotor flux is directly measured using search coils or approximated from terminal data. This position is crucial for precise orientation. However, utilizing sensors to gather flux information makes it difficult to use an induction machine that is already manufactured because the installation of such sensors can only be done during manufacturing. Figure 1.5 presents a schematic diagram for DFOC of DFIG.

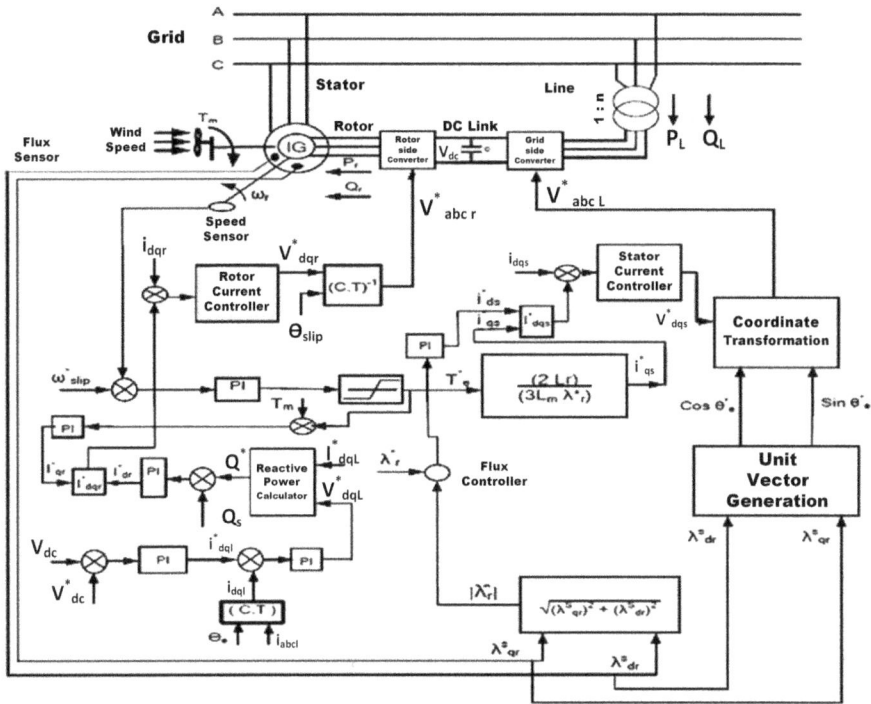

FIGURE 1.5 DFOC of a wind-driven DFIG.

1.4.2 IFOC OF A DFIG DRIVEN BY A WT

With the exception of how the flux position (θ_e) in DFOC is formed, IFOC mechanism is substantially the same as DFOC. To independently adjust the torque and rotor flux, the currents of the stator synchronously rotating vector components, i_{ds}^e and i_{qs}^e, are used. Rotor flux orientation may be demonstrated to typically provide true decoupling control. The slip and mechanical speed signals are used to construct the unit vector signal θ_e that converts the stator voltage vectors from synchronous frame into stationary frame. The scheme of IFOC is shown in Figure 1.6.

1.5 MODEL PREDICTIVE CONTROL (MPC) OF DFIG

The main distinction between the well-known feedback control schemes with PI controllers and the MPC formulation used to control AC drives is the pre-calculation of the behavior of this controlled system and the consideration of this behavior in the control signal before a difference between the real value and the reference value actually occurs. In contrast, the feedback control only reacts and attempts to correct a control difference after it has already appeared.

The MPC uses a cost function 'convergence condition', which replaces the PI regulators and current control loops of VOC and the hysteresis comparators used by

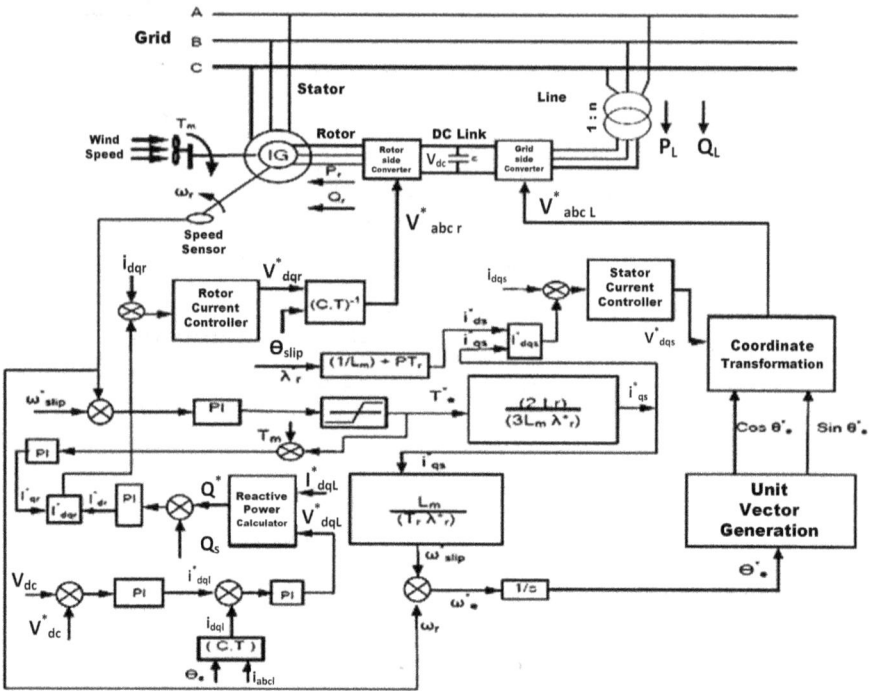

FIGURE 1.6   IFOC of a wind-driven DFIG.

DTC. This has contributed to achieving a quick dynamic response, producing a better steady-state performance and minimizing the complication and ripples as well [61–64]. The adopted cost function has two components of the absolute error of the variables that need to be controlled; it also has the merit of flexibility and ability to achieve various controller objectives. MPC is used in many applications; in [31], it is used to enhance the performance of the power system, and in [32], it is used with a microgrid system. PC has two topologies which are usually adopted with a DFIG: finite control set predictive control (FCS PC), which was adopted in [65–72], and continuous control set predictive control (CCS PC), which was adopted in [29, 73]. CCS PC has a fixed switching frequency, as it requires a PWM modulator so that the selected optimal voltages can be applied to the machine; meanwhile, FCS PC does not need a modulator as it chooses the voltages from specific vectors, usually eight vectors (six active and two null), which reduces the system's complexity, but its switching frequency is variable. FCS PC reduces the switching actions in comparison with CCS PC, resulting in a lower number of commutations. The control structures of FCS PC and CCS PC are presented in Figures 1.7 and 1.8, respectively, clarifying the difference between the two topologies. The vector $u(k)$ denotes the controlled variable, and $S(k)$ denotes the switching state, meanwhile $d(k)$ refers to the duty ratio.

The following strategy, illustrated in Figure 1.9, characterizes the methodology of the MPC technique:

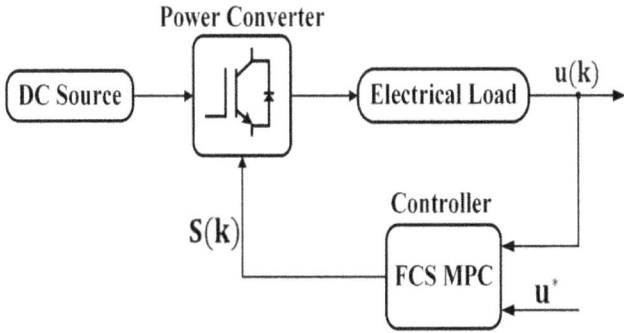

FIGURE 1.7 FCS PC structure.

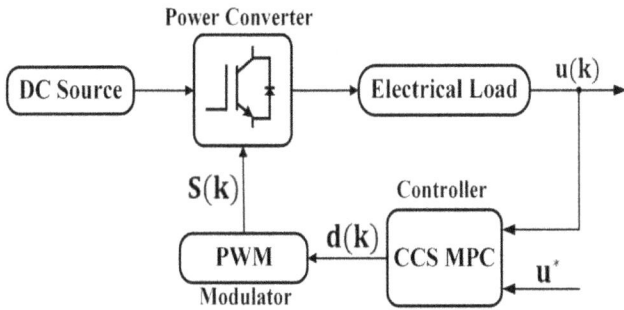

FIGURE 1.8 CCS PC structure.

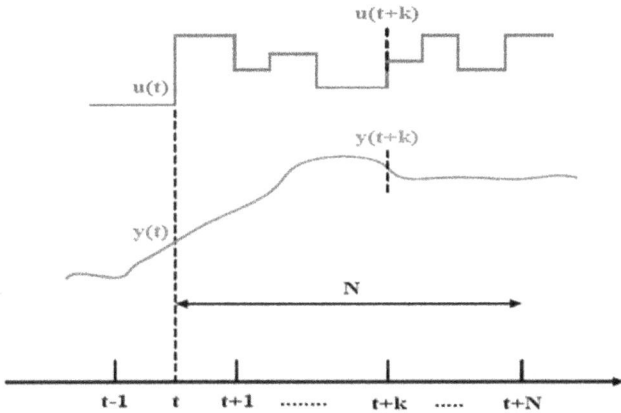

FIGURE 1.9 Methodology of MPC technique.

1. The process model is used to forecast the future outputs for a predetermined horizon N, also known as the prediction horizon, at each instant t. These anticipated results, denoted by $y(t+k)$ for $k=1 \ldots N$, depend on the values known as of instant t (previous inputs and outputs) as well as the upcoming control signals, denoted by $u(t+k)$, $k=0 \ldots N-1$, which are those to be calculated and transmitted to the system.

2. To maintain the process as closely as possible to the reference trajectory $R(t+k)$, the set of future control signals is calculated by optimizing a predetermined criterion (which can be the set point itself or a close approximation of it). A quadratic function of the errors between the expected output signal and the predicted reference trajectory often represents this requirement. Most of the time, the control effort is part of the cost function. If the criterion is quadratic, the model is linear, and there are no restrictions, an explicit solution can be found; otherwise, an iterative optimization method must be utilized. In other instances, assumptions are also made concerning the form of the future control rule, such as the assumption that it would be constant as of a particular instant [74].

3. Because $y(t+1)$ is already known at the next sampling moment, step 1 is performed with this new value, bringing all sequences up to date, and the control signal $u(t)$ is given to the process while the subsequent control signals calculated are rejected. Thus, the receding horizon idea is used to determine $u(t+1)$.

2 Modeling of the Wind Energy Conversion System Components

2.1 MATHEMATICAL MODEL OF DFIG

In the model of the DFIG shown in Figure 2.1, the stator voltage was chosen to be aligned with the direct axis of the rotating synchronous frame, so the frame in which the variables are expressed rotates with a speed equal to that of the stator voltage vector $(\omega_{\bar{u}s})$.

Here, the superscript 'sv' means that all variables are defined in the stator voltage frame which rotates with the synchronous speed $(\omega_{\bar{u}s})$; R_s, R_r, L_{ls}, L_{lr}, and L_m are the stator resistance, rotor resistance, stator leakage inductance, rotor leakage inductance, and mutual inductance, respectively; and ω_{me} is the electrical rotor angular speed and can be defined by $(\omega_{me,k} = p * \omega_{mech,k})$, where $\omega_{mech,k}$ refers to the mechanical rotor angular speed.

From Figure 2.1, and using a sampling time (T_s), the voltage balance in the stator and rotor can be represented at instant (KT_s) as follows:

$$\bar{u}_{s,k}^{sv} = R_s \bar{i}_{s,k}^{sv} + \frac{d\bar{\Psi}_{s,k}^{sv}}{dt} + j\omega_{\bar{u}s,k}\bar{\Psi}_{s,k}^{sv} \tag{2.1}$$

$$\bar{u}_{r,k}^{sv} = R_r \bar{i}_{r,k}^{sv} + \frac{d\bar{\Psi}_{r,k}^{sv}}{dt} + j\overbrace{\left(\omega_{\bar{u}s,k} - \omega_{me,k}\right)}^{\omega_{slip,k}}\bar{\Psi}_{r,k}^{sv} \tag{2.2}$$

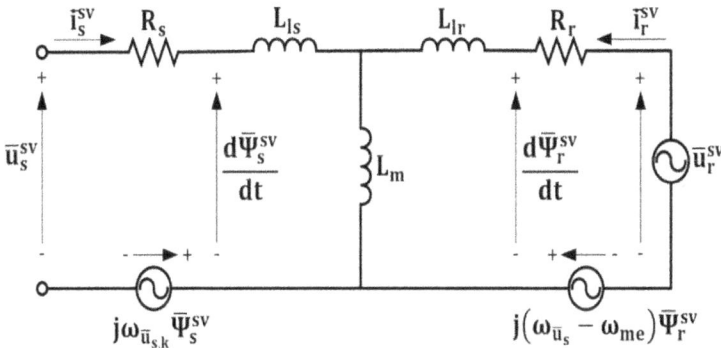

FIGURE 2.1 Per phase equivalent circuit model of DFIG.

DOI: 10.1201/9781003440529-3

Using the flux-current equations, the stator and rotor flux linkages can be expressed as follows:

$$\overline{\Psi}^{sv}_{s,k} = L_s \overline{i}^{sv}_{s,k} + L_m \overline{i}^{sv}_{r,k} \tag{2.3}$$

$$\overline{\Psi}^{sv}_{r,k} = L_r \overline{i}^{sv}_{r,k} + L_m \overline{i}^{sv}_{s,k} \tag{2.4}$$

where L_s and L_r are the stator and rotor inductances, respectively, which can be defined as

$$L_s = L_{ls} + L_m \tag{2.5}$$

$$L_r = L_{lr} + L_m \tag{2.6}$$

Using equations (2.1) and (2.2), the dynamic model of the DFIG in this coordinate system can be represented as follows:

$$\frac{d\overline{\Psi}^{sv}_{s,k}}{dt} = \overline{u}^{sv}_{s,k} - R_s \overline{i}^{sv}_{s,k} - j\omega_{\bar{u}s,k}\overline{\Psi}^{sv}_{s,k} \tag{2.7}$$

$$\frac{d\overline{\Psi}^{sv}_{r,k}}{dt} = \overline{u}^{sv}_{r,k} - R_r \overline{i}^{sv}_{r,k} - j\omega_{slip,k}\overline{\Psi}^{sv}_{r,k} \tag{2.8}$$

$$\frac{d\omega_{me,k}}{dt} = \frac{p}{J_g}\left(T_{me,k} - T_{d,k}\right) \tag{2.9}$$

where p and J_g are the number of pole pairs and inertia of the DFIG, respectively, and T_{me} and T_d are the mechanical torque applied to the shaft and the torque developed by the DFIG, respectively.

The torque developed by the DFIG can be defined by the following:

$$T_{d,k} = 1.5 pL_m \left(i^{sv}_{dr,k}i^{sv}_{qs,k} - i^{sv}_{qr,k}i^{sv}_{ds,k}\right) \tag{2.10}$$

After deriving the mathematical model of the DFIG, it can be used to construct the control schemes which are used to manage the dynamic performance of the DFIG.

2.2   MODELING OF THE WIND TURBINE

The wind turbine model is shown in Figure 2.2, in which the turbine drives a DFIG which supplies an isolated load, the overall system is defined as WECS. The speed of the DFIG is denoted by $\left(\omega_g \text{ or } \omega_{mech}\right)$, meanwhile the turbine speed is expressed as $\left(\omega_t\right)$. As known, we can manage the speed of the turbine via two methods: the first one is through managing β or by controlling T_g. There's an important ratio

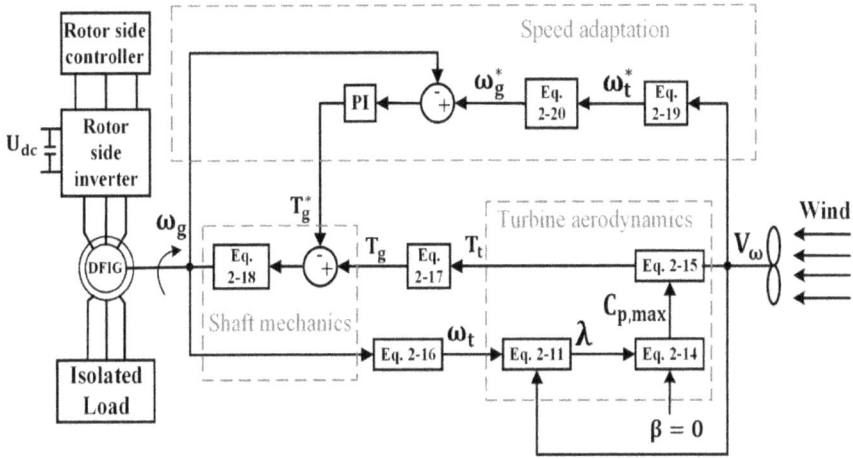

FIGURE 2.2 Model of wind turbine and speed adaptation mechanism.

used for evaluating the speed of the turbine, it's called as tip speed ratio (TSR) and can be expressed as:

$$\lambda = \frac{\omega_t R}{V_\omega} \tag{2.11}$$

The power of the wind (P_ω) can be represented by:

$$P_\omega = 0.5\rho A V_\omega^3 \tag{2.12}$$

The power of the turbine (P_t) can be expressed as:

$$P_t = C_p P_\omega = 0.5\rho A C_p V_\omega^3 \tag{2.13}$$

The power coefficient (C_p) can be represented in terms of (TSR), and (β) as follows:

$$C_p = \left[0.5 - 0.00167(\beta - 2)\right] \sin\left[\frac{\pi(\lambda + 0.1)}{10 - 0.3(\beta - 2)}\right] - 0.00184(\lambda - 3)(\beta - 2) \tag{2.14}$$

The turbine torque (T_t) can be calculated as follows:

$$T_t = \frac{P_t}{\omega_t} = \frac{0.5\rho A C_p V_\omega^3}{\omega_t} \tag{2.15}$$

There's an urgent need for using a gearbox ratio (G) to achieve the equilibrium among the turbine's shaft and the generator's shaft, so the speed of the generator's shaft is related to the turbine speed by the following relationship:

$$\omega_g = G\omega_t \tag{2.16}$$

The torque of the DFIG's shaft can be represented by:

$$T_g = \frac{T_t}{G}$$ (2.17)

We can represent the mechanical shaft by utilizing a two-mass model as follows:

$$T_t - GT_g - FG\omega_t = \left(\frac{J_t}{G} + GJ_g \right) \frac{d\omega_t}{dt}$$ (2.18)

The turbine must be managed to operate at an optimal value of TSR (λ_{opt}), to be able to extract the maximum available power; the reference values of turbine and generator speeds which can handle this condition can be represented by:

$$\omega_t^* = \frac{\lambda_{opt} V_\omega}{R}$$ (2.19)

$$\omega_g^* = G\omega_t^*$$ (2.20)

2.3   MATHEMATICAL MODEL OF THE ISOLATED LOAD UNDER STUDY

We used a three-phase SCIM as an isolated load which can be used for several applications like: water pumping, drilling machines, compressors, etc.; for this purpose, we should present a detailed mathematical model for the adopted IM. Figure 2.3 introduces the per phase equivalent circuit of the IM, which is represented in the stationary reference frame $(\alpha^s - \beta^s)$.

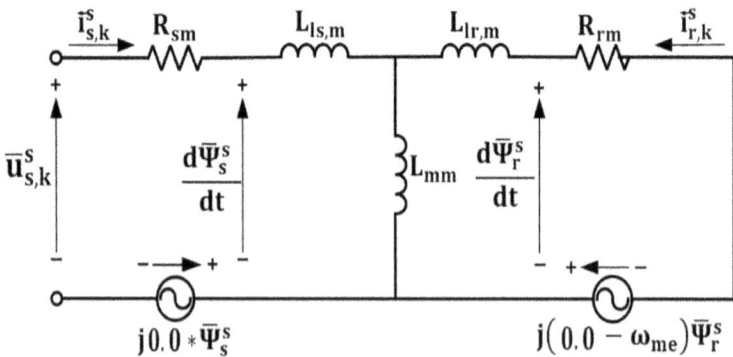

FIGURE 2.3   Per phase equivalent circuit model of IM.

The voltage balance equations in Figure 2.3 can be represented by:

$$\bar{u}_{s,k}^s = R_{sm}\bar{i}_{s,k}^s + \frac{d\bar{\Psi}_{s,k}^s}{dt} \tag{2.21}$$

$$0.0 = R_{rm}\bar{i}_{r,k}^{sv} + \frac{d\bar{\Psi}_{r,k}^s}{dt} - j\omega_{m,k}\bar{\Psi}_{r,k}^s \tag{2.22}$$

Now, the dynamics of the IM can be easily described as follows:

First, the derivative of the stator current vector can be obtained using the stator voltage equation (2.21)

$$\frac{d\bar{i}_{s,k}^s}{dt} = \frac{L_{rm}L_{mm}}{L_{sm}L_{rm}^2 + L_{rm}L_{mm}^2}\left(\frac{L_{rm}}{L_{mm}}\bar{u}_{s,k}^s - \frac{R_{sm}L_{rm}}{L_{mm}}\bar{i}_{s,k}^s + R_{rm}\bar{i}_{r,k}^s - j\omega_{m,k}\bar{\Psi}_{r,k}^s\right) \tag{2.23}$$

After that, from equation (2.22), we can obtain:

$$\frac{d\bar{\Psi}_{r,k}^s}{dt} = -R_{rm}\bar{i}_{r,k}^s + j\omega_{m,k}\bar{\Psi}_{r,k}^s \tag{2.24}$$

$$\frac{d\omega_{m,k}}{dt} = \frac{p_{IM}}{J_{IM}}\left(T_{e,k} - T_{l,k}\right) \tag{2.25}$$

where $\omega_{m,k}$ is the electrical angular speed of the IM and can be defined by $\left(\omega_{m,k} = p_{IM} * \omega_{mech,k_{IM}}\right)$, where $\omega_{mech,k_{IM}}$ refers to the mechanical angular speed of the IM, $T_{l,k}$ refers to the load torque; $T_{e,k}$ denotes the electromagnetic torque developed by the IM and can be defined by:

$$T_{e,k} = 1.5p_{IM}\frac{L_{mm}}{L_{rm}}\left(\Psi_{dr,k}^s i_{qs,k}^s - \Psi_{qr,k}^s i_{ds,k}^s\right) \tag{2.26}$$

The control topology which is implemented for controlling the IM is the MPDTC, and it will be described in a detailed manner in Chapter 3, Section 3.3.

2.4 MODELING OF ROTOR SIDE CONVERTER (VSI)

To apply the reference voltage signals to the rotor terminals, a two-level three phase VSI is utilized. The rotor VSI has 2^3 switching combinations, which provides eight possible voltage vectors (six active and two nulls). The structure of the inverter and the voltages space representation are shown in Figure 2.4.

The possible switching actions of the inverter can be mathematically represented using a single function as follows:

$$S = \frac{2}{3}\left(S_1 + \alpha S_2 + \alpha^2 S_3\right) \tag{2.27}$$

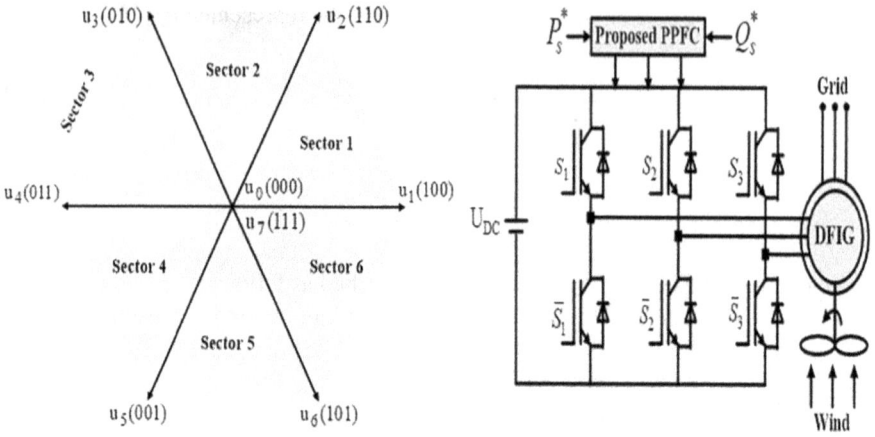

FIGURE 2.4   Structure and voltages representation of rotor VSI.

where $\alpha = e^{j\frac{2\pi}{3}}$, and $S_{i,1,2,3}$ and $\overline{S}_{i,1,2,3}$ represent the on and off switching actions of the upper and lower switches and they equal $\{1,0\}$ and $\{0,1\}$, respectively.

Using the switching state S, then the rotor voltage is calculated in terms of the DC link voltage U_{dc} by

$$\overline{u}_r = SU_{dc} \qquad\qquad (2.28)$$

3 Analyzing and Enhancing the Performance of a Standalone Variable Speed DFIG

3.1 CONTROL SCHEMES ADOPTED FOR THE STANDALONE DFIG

3.1.1 Stator Voltage-Oriented Control (SVOC) Technique

VOC is a strategy that depends in its operation on decomposing the stator current into two components: the torque component, which produces the torque, and the magnetizing component, which produces the flux. Thereafter, the two components are decoupled and individually controlled [75–79].

In [80], the SVOC principle is adopted, in which the stator voltage vector $\bar{u}_{s,k}$ is set in the direction of the d-axis of the synchronous frame, which rotates with a speed of $\omega_{\bar{u}s,k}$, as shown in Figure 3.1, so there is no voltage along the q-axis. As noted in Figure 3.1, there are three reference frames: the stationary reference frame $\left(\alpha^s - \beta^s\right)$, the rotor reference frame $\left(d^r - q^r\right)$, and the synchronous rotating frame $\left(d^{\bar{u}_s} - q^{\bar{u}_s}\right)$.

Thus, by considering the orientations presented in Figure 3.1, the following relations can be concluded under SVOC:

$$u_{ds,k}^{sv} = \left|\bar{u}_{s,k}^{sv}\right| \quad \text{and} \quad u_{qs,k}^{sv} = 0 \tag{3.1}$$

Equation (3.1) can be represented in the d–q axes as follows:

$$u_{ds,k}^{sv} = R_s i_{ds,k}^{sv} + \frac{d\Psi_{ds,k}^{sv}}{dt} - \omega_{\bar{u}s,k} \Psi_{qs,k}^{sv} \tag{3.2}$$

$$u_{qs,k}^{sv} = R_s i_{qs,k}^{sv} + \frac{d\Psi_{qs,k}^{sv}}{dt} + \omega_{\bar{u}s,k} \Psi_{ds,k}^{sv} \tag{3.3}$$

Using equations (3.2) and (3.3), and taking into consideration that the stator voltage drop on the resistance R_s is very small compared with the other terms and can be neglected, it is found that $\dfrac{d\bar{\Psi}_{s,k}^{sv}}{dt} \cong 0.0$ under steady-state operation, and the following relationships can be derived:

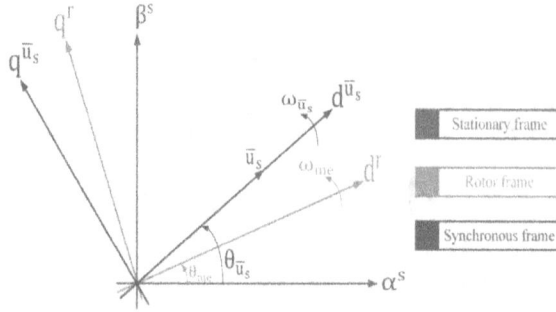

FIGURE 3.1   Relationships between reference frames.

$$\Psi_{ds,k}^{sv} \approx \frac{u_{qs,k}^{sv}}{\omega_{\bar{u}s,k}} \approx 0.0 \quad \text{and} \quad \Psi_{qs,k}^{sv} \approx \left|\bar{\Psi}_{s,k}^{sv}\right| \approx -\frac{u_{ds,k}^{sv}}{\omega_{\bar{u}s,k}} \tag{3.4}$$

Using equation (3.4), equation (2.3) can be represented in the $d-q$ axes as follows:

$$\Psi_{ds,k}^{sv} = L_s i_{ds,k}^{sv} + L_m i_{dr,k}^{sv} \approx 0.0 \tag{3.5}$$

$$\Psi_{qs,k}^{sv} = L_s i_{qs,k}^{sv} + L_m i_{qr,k}^{sv} \approx -\frac{u_{ds,k}^{sv}}{\omega_{\bar{u}s,k}} \tag{3.6}$$

From equations (3.5) and (3.6), the following can be concluded:

$$i_{ds,k}^{sv} = -\left(\frac{L_m}{L_s}\right) i_{dr,k}^{sv} \tag{3.7}$$

$$i_{qs,k}^{sv} = \frac{\Psi_{qs,k}^{sv}}{L_s} - \left(\frac{L_m}{L_s}\right) i_{qr,k}^{sv} = -\left(\frac{L_m}{L_s}\right) i_{qr,k}^{sv} - \frac{u_{ds,k}^{sv}}{\omega_{\bar{u}s,k} L_s} \tag{3.8}$$

Using equations (3.1), (3.7), and (3.8), the active power $\left(P_{s,k}\right)$ and reactive power $\left(Q_{s,k}\right)$ of the DFIG under SVOC can be calculated as follows:

$$P_{s,k} = 1.5\left(u_{ds,k}^{sv} i_{ds}^{sv} + u_{qs,k}^{sv} i_{qs}^{sv}\right) = 1.5 u_{ds,k}^{sv} i_{ds}^{sv} = -1.5\frac{L_m}{L_s} u_{ds,k}^{sv} i_{dr,k}^{sv} \tag{3.9}$$

$$Q_{s,k} = 1.5\left(u_{qs,k}^{sv} i_{ds}^{sv} - u_{ds,k}^{sv} i_{qs}^{sv}\right) = -1.5 u_{ds,k}^{sv} i_{qs}^{sv} = 1.5\left[\frac{L_m}{L_s} u_{ds,k}^{sv} i_{qr,k}^{sv} + \frac{\left(u_{ds,k}^{sv}\right)^2}{\omega_{\bar{u}s,k} L_s}\right] \tag{3.10}$$

$$= 1.5\frac{L_m}{L_s} u_{ds,k}^{sv} \left(i_{qr,k}^{sv} + \frac{u_{ds,k}^{sv}}{\omega_{\bar{u}s,k} L_m}\right)$$

Furthermore, the developed torque $T_{d,k}$ can be reformulated by substituting from equations (3.7) and (3.8) into (2.10) as follows:

$$T_{d,k} = 1.5pL_m \left[i_{dr,k}^{sv} \left(-\frac{L_m}{L_s} i_{qr,k}^{sv} - \frac{u_{ds,k}^{sv}}{\omega_{\bar{u}_{s,k}} L_s} \right) + i_{qr,k}^{sv} \left(\frac{L_m}{L_s} \right) i_{dr,k}^{sv} \right]$$

$$= 1.5p \frac{L_m}{L_s} \left(-L_m i_{dr,k}^{sv} i_{qr,k}^{sv} - \frac{u_{ds,k}^{sv}}{\omega_{\bar{u}_{s,k}}} i_{dr,k}^{sv} + L_m i_{dr,k}^{sv} i_{qr,k}^{sv} \right) \qquad (3.11)$$

$$= -1.5p \frac{L_m}{L_s \omega_{\bar{u}_{s,k}}} u_{ds,k}^{sv} i_{dr,k}^{sv}$$

Equations (3.–9)–(3.11) show that the active power and developed torque can be controlled by adjusting the torque current component $i_{dr,k}^{sv}$, while the reactive power can be controlled by adjusting the excitation current component $i_{qr,k}^{sv}$.

Equation (2.4) can be represented in the d–q axes as follows:

$$\Psi_{dr,k}^{sv} = L_r i_{dr,k}^{sv} + L_m i_{ds,k}^{sv} \qquad (3.12)$$

$$\Psi_{qr,k}^{sv} = L_r i_{qr,k}^{sv} + L_m i_{qs,k}^{sv} \qquad (3.13)$$

Using equation (3.7), equation (3.12) can be represented as follows:

$$\Psi_{dr,k}^{sv} = L_r i_{dr,k}^{sv} + L_m \left(-\frac{L_m}{L_s} i_{dr,k}^{sv} \right) = \frac{L_s L_r - L_m^2}{L_s} i_{dr,k}^{sv} = \sigma L_r i_{dr,k}^{sv} \qquad (3.14)$$

where $\left(\sigma = \frac{L_s L_r - L_m^2}{L_s L_r} = 1 - \frac{L_m^2}{L_s L_r} \right)$ is the leakage factor.

Equation (3.14) can be reformulated in a general case as follows:

$$\Psi_{dr,k}^{sv} = \frac{L_m}{L_s} \Psi_{ds,k}^{sv} + \sigma L_r i_{dr,k}^{sv} \qquad (3.15)$$

Using equation (3.8), equation (3.13) can be represented as follows:

$$\Psi_{qr,k}^{sv} = L_r i_{qr,k}^{sv} + L_m \left(\frac{\Psi_{qs,k}^{sv}}{L_s} - \frac{L_m}{L_s} i_{qr,k}^{sv} \right) = \frac{L_m}{L_s} \Psi_{qs,k}^{sv} + \frac{L_s L_r - L_m^2}{L_s} i_{qr,k}^{sv}$$

$$\qquad (3.16)$$

$$= \frac{L_m}{L_s} \Psi_{qs,k}^{sv} + \sigma L_r i_{qr,k}^{sv}$$

Equation (2.2) can be represented in the d–q axes as follows:

$$u_{dr,k}^{sv} = R_r i_{dr,k}^{sv} + \frac{d\Psi_{dr,k}^{sv}}{dt} - \omega_{slip,k} \Psi_{qr,k}^{sv} \qquad (3.17)$$

$$u_{qr,k}^{sv} = R_r i_{qr,k}^{sv} + \frac{d\Psi_{qr,k}^{sv}}{dt} + \omega_{slip,k}\Psi_{dr,k}^{sv} \tag{3.18}$$

Using equations (3.15) and (3.16), knowing that $\left(\dfrac{d\bar{\Psi}_{s,k}^{sv}}{dt} \cong 0.0\right)$ under-steady state operation, equations (3.17) and (3.18) can be reformulated as follows:

$$u_{dr,k}^{sv} = \underbrace{R_r i_{dr,k}^{sv} + \sigma L_r \frac{di_{dr,k}^{sv}}{dt}}_{u'_{dr,k}\ =\ Active\ term} - \underbrace{\omega_{slip,k}\left(\frac{L_m}{L_s}\Psi_{qs,k}^{sv} + \sigma L_r i_{qr,k}^{sv}\right)}_{\Delta u_{dr,k}^{sv}\ =\ Compensation\ term} \tag{3.19}$$

$$u_{qr,k}^{sv} = R_r i_{qr,k}^{sv} + \frac{L_m}{L_s}\frac{d\Psi_{qs,k}^{sv}}{dt} + \sigma L_r \frac{di_{qr,k}^{sv}}{dt} + \omega_{slip,k}\sigma L_r i_{dr,k}^{sv}$$

$$= \underbrace{R_r i_{qr,k}^{sv} + \sigma L_r \frac{di_{qr,k}^{sv}}{dt}}_{u'_{qr,k}\ =\ Active\ term} + \underbrace{\omega_{slip,k}\sigma L_r i_{dr,k}^{sv}}_{\Delta u_{qr,k}^{sv}\ =\ Compensation\ term} \tag{3.20}$$

From equations (3.19) and (3.20), the derivatives of the stator and rotor current components to be used to formulate the adopted control structures can be found as follows [81]:

$$\frac{di_{dr,k}^{sv}}{dt} = \frac{L_m^2 + L_r L_t}{L_r^2 L_t}\left[u_{dr,k}^{sv} - R_r i_{dr,k}^{sv} + \frac{L_r}{L_m}\omega_{slip,k}\left(\Psi_{qs,k}^{sv} - L_t i_{qs,k}^{sv}\right)\right]$$

$$- \frac{L_m}{L_r L_t}\left(u_{ds,k}^{sv} - R_s i_{ds,k}^{sv} + \omega_{\bar{u}s,k}\Psi_{qs,k}^{sv}\right) \tag{3.21}$$

$$\frac{di_{qr,k}^{sv}}{dt} = \frac{L_m^2 + L_r L_t}{L_r^2 L_t}\left[u_{qr,k}^{sv} - R_r i_{qr,k}^{sv} - \frac{L_r}{L_m}\omega_{slip,k}\left(\Psi_{ds,k}^{sv} - L_t i_{ds,k}^{sv}\right)\right]$$

$$- \frac{L_m}{L_r L_t}\left(u_{qs,k}^{sv} - R_s i_{qs,k}^{sv} - \omega_{\bar{u}s,k}\Psi_{ds,k}^{sv}\right) \tag{3.22}$$

where $\left(L_t = \dfrac{L_s L_r - L_m^2}{L_r} = L_s - \dfrac{L_m^2}{L_r} = \sigma L_s\right)$ is the stator transient inductance, the stator flux components $\left(\Psi_{ds,k}^{sv}\ \text{and}\ \Psi_{qs,k}^{sv}\right)$ can be calculated using equations (3.5) and (3.6). The derivative of the stator current components can be evaluated in the same manner.

Equations (3.19) and (3.20) formulate the base operation of SVOC, as the reference d–q rotor voltage components are obtained by summing the active voltage terms provided by the proportional–integral (PI) current controllers and the compensating terms calculated in terms of generator variables.

The schematic diagram of the SVOC is shown in Figure 3.2, in which the reference rotor current components $\left(i_{dr,k}^{*}\ \text{and}\ i_{qr,k}^{*}\right)$ are obtained using the errors of the load active power and load voltage with the aid of two PI regulators to provide the

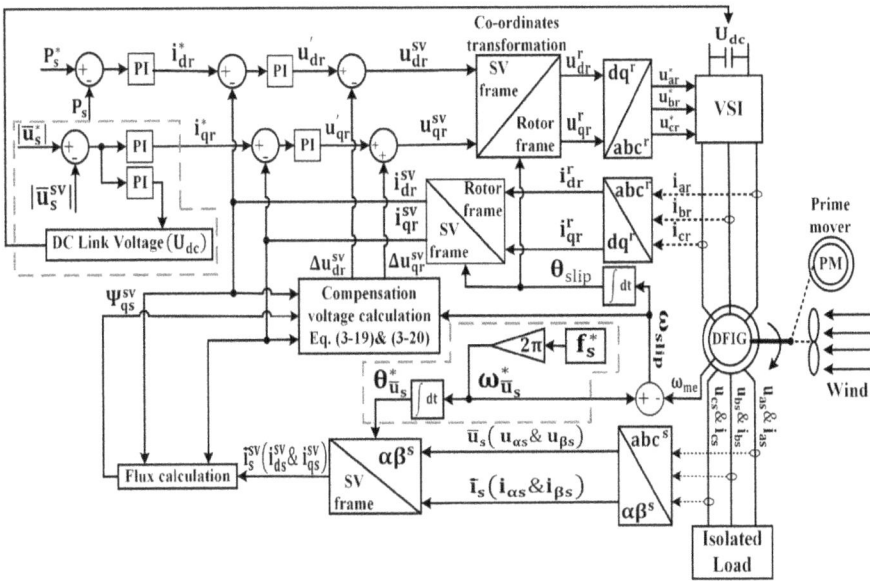

FIGURE 3.2 Scheme of SVOC approach for the DFIG.

required load power and maintain a constant load voltage all time which is considered as an important requirement for standalone systems. After that, the rotor current errors are fed to the PI current regulators to generate the active terms of rotor voltage $u'_{dr,k}$ and $u'_{qr,k}$. Also, it's necessary to keep the frequency of the load voltage constant, so, the reference frequency (f^*_s) is utilized to calculate the reference synchronous angular speed $(\omega^*_{\bar{u}_s})$ and also, the stator voltage angle $(\theta^*_{\bar{u}_s})$, which is required for the coordinates' transformation and further used in calculating the slip frequency, which is needed for calculating the compensation voltage components $\Delta u^{sv}_{dr,k}$ and $\Delta u^{sv}_{qr,k}$. Finally, the rotor voltage components $u^{sv}_{dr,k}$ and $u^{sv}_{qr,k}$ can be easily obtained after adding the compensation terms to the active terms as stated in equations (3.19) and (3.20).

3.1.2 Model Predictive Current Control (MPCC) Technique

MPCC utilizes a simple cost function consisting of two similar terms, which are the errors between the actual and reference values of the rotor current, so it does not need a weighting factor. The actual rotor current components $\left(i^{sv}_{dr,k+1}\right.$ and $\left.i^{sv}_{qr,k+1}\right)$ can be predicted using Taylor expansion, while the reference values of the rotor current $i^*_{dr,k+1}$ and $i^*_{qr,k+1}$ can be directly obtained using the errors of the load active power and load voltage with the aid of two PI regulators as mentioned previously.

The scheme of MPCC can be constructed as shown in Figure 3.3. First, the stator voltage, stator current, and rotor current are measured and then sampled. The rotor speed $(\omega_{me,k})$ is measured and then integrated to find the rotor position $(\theta_{me,k})$. The rotor position can be estimated at instant $(k+1)T_s$ using the following formulation:

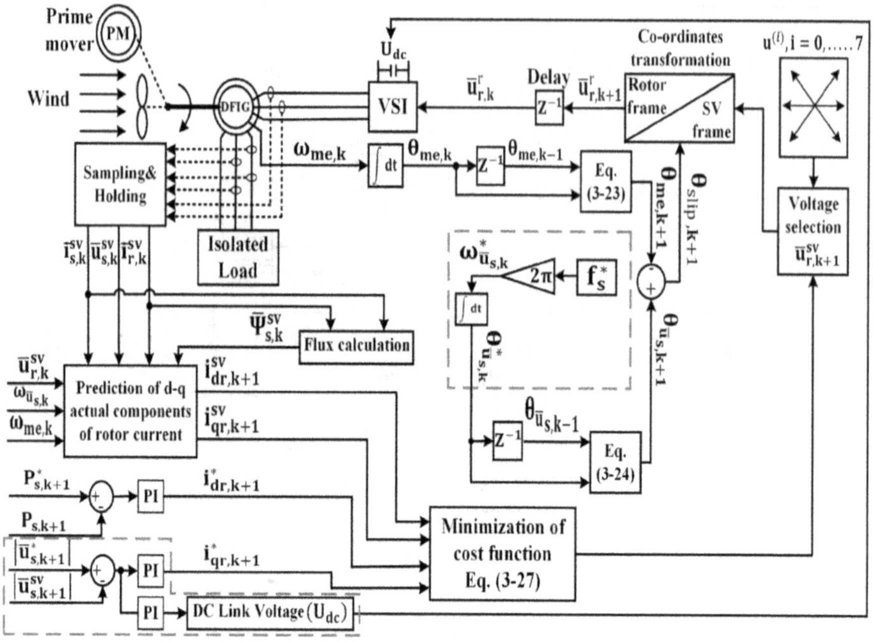

FIGURE 3.3   Scheme of MPCC approach for the DFIG.

$$\theta_{me,k+1} = \theta_{me,k} + \left(\frac{\theta_{me,k} - \theta_{me,k-1}}{\Delta T} \right) T_s \qquad (3.23)$$

In the same manner, the reference stator voltage angle $\left(\theta^*_{\bar{u}s,k} \right)$ can be evaluated by integrating the reference synchronous angular speed $\left(\omega^*_{\bar{u}s} \right)$, and then estimated at instant $(k+1)T_s$ using the following formulation:

$$\theta_{\bar{u}s,k+1} = \theta_{\bar{u}s,k} + \left(\frac{\theta_{\bar{u}s,k} - \theta_{\bar{u}s,k-1}}{\Delta T} \right) T_s \qquad (3.24)$$

Prediction of the actual values of the rotor current can be implemented using Taylor expansion as follows:

$$i^{sv}_{dr,k+1} = i^{sv}_{dr,k} + \left(\frac{di^{sv}_{dr,k}}{dt} \right) T_s \qquad (3.25)$$

$$i^{sv}_{qr,k+1} = i^{sv}_{qr,k} + \left(\frac{di^{sv}_{qr,k}}{dt} \right) T_s \qquad (3.26)$$

The derivative components of the rotor current can be calculated using equations (3–21) and (3–22), which prove that the cost function of MPCC depends on the machine parameters as it includes estimated variables.

Finally, the actual and reference rotor current components are fed to the cost function, which can be expressed as follows:

$$\Gamma_i = \left| i^*_{dr,k+1} - i^{sv}_{dr,k+1} \right|_i + \left| i^*_{qr,k+1} - i^{sv}_{qr,k+1} \right|_i \tag{3.27}$$

where the subscript (i) denotes the sectors $(0, \ldots, 7)$.

The cost function (3.27) represents the core of the MPCC, based upon which the optimal voltage vectors are selected.

3.1.3 Model Predictive Direct Torque Control (MPDTC) Technique

MPDTC depends in its operation on regulating the torque and rotor flux [82, 83], which can be accomplished by eliminating the difference between the reference and predicted actual torque signals $\left(T^*_{d,k+1} \text{ and } T_{d,k+1} \right)$ and between the reference and predicted actual rotor flux signals $\left(\overline{\Psi}^*_{r,k+1} \text{ and } \overline{\Psi}^{sv}_{r,k+1} \right)$; thus, the cost function can be expressed as follows:

$$\wedge_i = \left| T^*_{d,k+1} - T_{d,k+1} \right|_i + \omega_f \left\| \overline{\Psi}^*_{r,k+1} \right| - \left| \overline{\Psi}^{sv}_{r,k+1} \right\|_i \tag{3.28}$$

Equation (3.28) includes variables that need to be evaluated using the machine parameters; it also requires current calculation, which is used for estimating and predicting the torque and rotor flux, and all of this increases the computation time. Furthermore, the function needs a weighting value $\left(\omega_f \right)$, that is required for ensuring the equilibrium among the torque and rotor flux. If the weighting factor is inappropriately selected, the torque and rotor flux ripples are negatively affected. As mentioned in [82], there is a procedure which tried to choose the optimal weighting factor in an online manner, which reduced the deviations in the torque and rotor flux, but unfortunately, the computational burden was affected. Hence, the weighting factor is still a major obstacle to MPDTC.

The reference value of rotor flux $\left| \overline{\Psi}^*_{r,k+1} \right|$ can be evaluated as follows:

$$\left| \overline{\Psi}^*_{r,k+1} \right| = \sqrt{\left(\Psi^*_{dr,k+1} \right)^2 + \left(\Psi^*_{qr,k+1} \right)^2} \tag{3.29}$$

where $\Psi^*_{dr,k+1}$ and $\Psi^*_{qr,k+1}$ can be calculated as follows:

$$\Psi^*_{dr,k+1} = L_r i^*_{dr,k+1} + L_m i^*_{ds,k+1} \tag{3.30}$$

$$\Psi^*_{qr,k+1} = L_r i^*_{qr,k+1} + L_m i^*_{qs,k+1} \tag{3.31}$$

where $i^*_{dr,k+1}$ and $i^*_{qr,k+1}$ are directly can be found directly by utilizing the errors of the load active power and load voltage with the aid of two PI regulators as stated previously, while $i^*_{ds,k+1}$ and $i^*_{qs,k+1}$ can be expressed in terms of the reference components of the rotor current, as follows:

$$i^*_{ds,k+1} = -\left(\frac{L_m}{L_s}\right)i^*_{dr,k+1} \tag{3.32}$$

$$i^*_{qs,k+1} = -\left(\frac{L_m}{L_s}\right)i^*_{qr,k+1} - \frac{u^{sv}_{ds,k+1}}{\omega_{\bar{u}s,k+1}L_s} \tag{3.33}$$

The reference value of the torque at instant $(k+1)T_s$ can be represented as follows:

$$T^*_{d,k+1} = 1.5pL_m\left(i^*_{dr,k}i^*_{qs,k} - i^*_{qr,k}i^*_{ds,k}\right) \tag{3.34}$$

The actual value of the rotor flux $\left|\Psi^{sv}_{r,k+1}\right|$ can be predicted and calculated at instant $(k+1)T_s$ in the following manner:

$$\left|\bar{\Psi}^{sv}_{r,k+1}\right| = \sqrt{\left(\Psi^{sv}_{dr,k+1}\right)^2 + \left(\Psi^{sv}_{qr,k+1}\right)^2} \tag{3.35}$$

where $\Psi^{sv}_{dr,k+1}$ and $\Psi^{sv}_{qr,k+1}$ can be formulated as follows:

$$\Psi^{sv}_{dr,k+1} = L_r i^{sv}_{dr,k+1} + L_m i^{sv}_{ds,k+1} \tag{3.36}$$

$$\Psi^{sv}_{qr,k+1} = L_r i^{sv}_{qr,k+1} + L_m i^{sv}_{qs,k+1} \tag{3.37}$$

where the rotor current components $i^{sv}_{dr,k+1}$ and $i^{sv}_{qr,k+1}$ can be predicted using equations (3.25) and (3.26), and the stator current components $i^{sv}_{ds,k+1}$ and $i^{sv}_{qs,k+1}$ can be predicted in the same manner or using the following formulations:

$$i^{sv}_{ds,k+1} = -\left(\frac{L_m}{L_s}\right)i^{sv}_{dr,k+1} \tag{3.38}$$

$$i^{sv}_{qs,k+1} = -\left(\frac{L_m}{L_s}\right)i^{sv}_{qr,k+1} - \frac{u^{sv}_{ds,k+1}}{\omega_{\bar{u}s,k+1}L_s} \tag{3.39}$$

The actual predicted value of the torque $T_{d,k+1}$ can be expressed as follows:

$$T_{d,k+1} = 1.5pL_m\left(i^{sv}_{dr,k+1}i^{sv}_{qs,k+1} - i^{sv}_{qr,k+1}i^{sv}_{ds,k+1}\right) \tag{3.40}$$

The scheme of MPDTC is illustrated in Figure 3.4, in which the stator voltage, stator current, and rotor current are measured and then sampled. The rotor speed $(\omega_{me,k})$ is measured and then integrated to find the rotor position $(\theta_{me,k})$, and then, the rotor position can be estimated at instant $(k+1)T_s$ as stated in equation (3.23). The stator voltage angle $(\theta^*_{\bar{u}s,k})$ can be estimated at instant $(k+1)T_s$ as stated in equation (3.24). The actual values of the stator and rotor currents are predicted using Taylor expansion, and these components are then used to calculate the actual values of the torque and rotor flux. The reference components of the rotor current are calculated using the

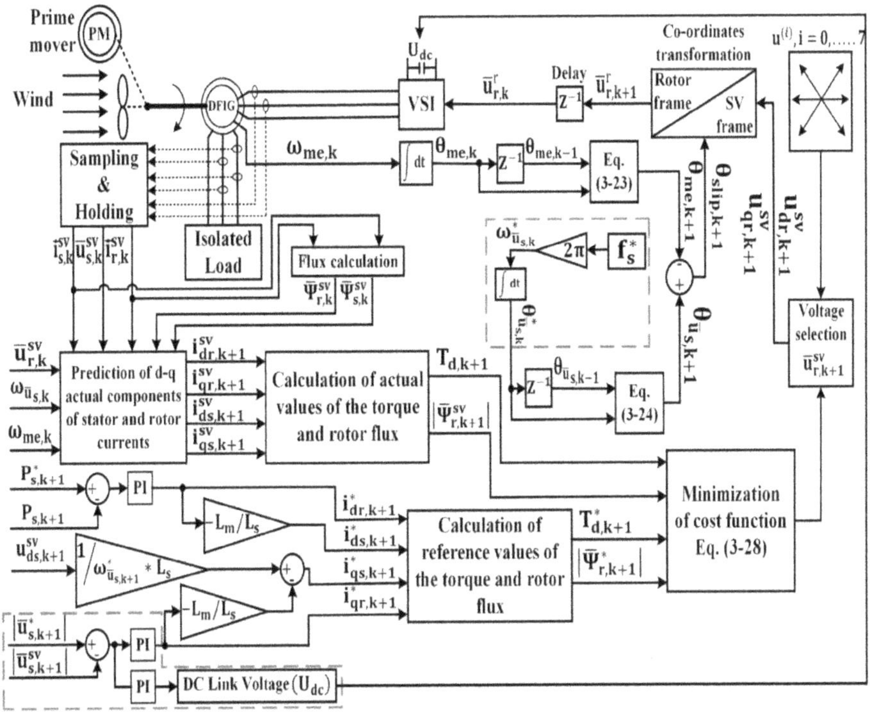

FIGURE 3.4 Scheme of MPDTC approach for the DFIG.

errors of the load active power and load voltage with the aid of two PI regulators as mentioned previously, then the reference components of the stator current can be easily obtained; after that, these components are used to compute the reference values of the torque and rotor flux. Finally, the reference and actual predicted values of the torque and rotor flux are fed to the cost function in equation (3.28).

3.1.4 PROPOSED PREDICTIVE VOLTAGE CONTROL (PVC) TECHNIQUE

The proposed PVC utilizes a very simple cost function that has two similar terms, which are the errors between the reference and actual values of the rotor voltage. The adopted cost function can be expressed as:

$$C_i = \left| u_{dr,k+1}^* - u_{dr,k+1}^{sv} \right|_i + \left| u_{qr,k+1}^* - u_{qr,k+1}^{sv} \right|_i \qquad (3.41)$$

As it's clear from (3.41), the adopted cost function is complication free; as its function is to minimize the error value between the reference and predicted actual values of the d-q components of the rotor voltage, so its components are analogous, thus it doesn't require a weighting factor which can cause a problem of mismatch as in case of MPDTC. Furthermore, the cost function is free of estimated variables which are obtained through the model parameters, this leads to handling the issue of system uncertainties and making it robust against parameters' variations.

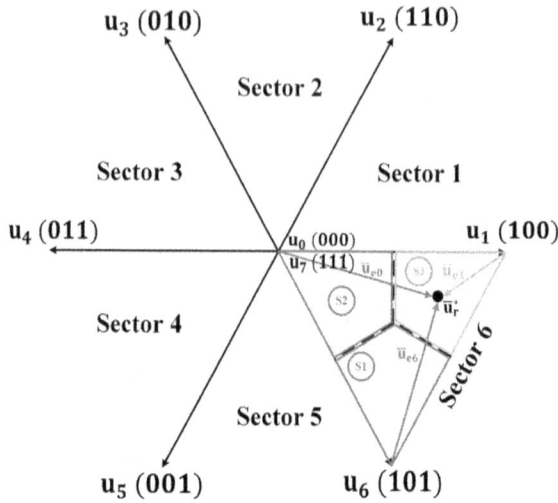

FIGURE 3.5 Mechanism of selection the optimal rotor voltage using PVC.

Figure 3.5 presents the mechanism by which the PVC can generate the rotor voltages which match with the required reference values, the main function of this mechanism is minimizing the error value between the reference and actual values of the rotor voltage. The possible behaviors of the controller are shown in Figure 3.5. The reference rotor voltage vector $\left(\bar{u}_r^*\right)$ is assumed to be present in sector 6, as shown in Figure 3.5. The cost function (3.41) starts checking which voltage vector can achieve minimum error, i.e., minimum deviation from the reference value. There're three bisectors created within sector 6, they are created by cutting the sector's inter-median lines in half ($S1$, $S2$, and $S3$). We can note that there are three vectors who can take on this position, which are $u_1\left(1\,0\,0\right)$, $u_6\left(1\,0\,1\right)$ and $u_0\left(0\,0\,0\right)$ or $u_7\left(1\,1\,1\right)$, but the switching state $\left\{u_6\left(1\,0\,1\right)\right\}$ is the best vector, since it results in the least value of deviation $\left(\bar{u}_{e1}\right)$. Consequently, the proposed PVC can achieve the control targets with a low computational burden by tracking the location of the reference vector and employing (3.41).

Now, the actual values of the predicted rotor voltage $\left(u_{dr,k+1}^{sv} \text{ and } u_{qr,k+1}^{sv}\right)$ can be obtained directly through the switching states of the VSI, i.e., utilizing the FCS principle, which evaluates the voltage vectors in terms of the inverter switching states without needing for PWM. To prove and confirm the validation and flexibility of our proposed predictive controller, the reference values of the rotor voltage $\left(u_{dr,k+1}^* \text{ and } u_{qr,k+1}^*\right)$ are generated in two different schemes: one of them is used with direct driven operation of the standalone DFIG; and the other is utilized with wind driven operation. The two methods will be discussed in a systematic manner in the following sections.

3.1.4.1 Proposed PVC Scheme under Direct-Driven Operation

The reference rotor voltage components $\left(u^*_{dr,k+1}\right)$ and $\left(u^*_{qr,k+1}\right)$ can be calculated using the errors of the torque and air-gap energy with the aid of two PI regulators, which are designed as following:

As mentioned in [82], the torque derivative $\left(T_{d,k+1}\right)$ can be represented by:

$$
\frac{dT_{d,k+1}}{dt} = -1.5p\frac{L_m}{\sigma L_s L_r}\left(\frac{L_m}{L_r}\Psi^{sv}_{qr,k+1}u^{sv}_{dr,k} + \Psi^{sv}_{dr,k+1}u^{sv}_{ds,k+1}\right)
$$
$$
-\left(\frac{R_s}{\sigma L_s} + \frac{R_r}{\sigma L_r}\right)T_{d,k+1} - \frac{L_m}{L_r}\omega_{me,k+1}\left(\Psi^{sv}_{qr,k+1}\right)^2
\tag{3.42}
$$

The air-gap energy $\left(E_{k+1}\right)$ and its derivative $\left(\dfrac{dE_{e,k+1}}{dt}\right)$ can be formulated as:

$$
E_{k+1} = 1.5p\frac{L_m}{\sigma L_s L_r}\left(\Psi^{sv}_{ds,k+1}\Psi^{sv}_{dr,k+1} + \Psi^{sv}_{qs,k+1}\Psi^{sv}_{qr,k+1}\right)
\tag{3.43}
$$

$$
\frac{dE_{e,k+1}}{dt} = 1.5p\frac{L_m}{\sigma L_s L_r}\left[
\begin{array}{l}
\Psi^{sv}_{qs,k+1}u^{sv}_{qr,k+1} - \left(\dfrac{R_s}{\sigma L_s} + \dfrac{R_r}{\sigma L_r}\right)\Psi^{sv}_{qs,k+1}\Psi^{sv}_{qr,k+1} \\[2mm]
+\dfrac{R_s L_m}{\sigma L_s L_r}\left(\left(\Psi^{sv}_{qs,k+1}\right)^2 + \left(\Psi^{sv}_{qr,k+1}\right)^2\right) \\[2mm]
-\omega_{slip,k+1}\Psi^{sv}_{qs,k+1}\Psi^{sv}_{dr,k+1}
\end{array}
\right]
\tag{3.44}
$$

The d-axis component of the rotor flux can be represented by:

$$
\Psi^{sv}_{dr,k} = \frac{L_m}{L_s}\Psi^{sv}_{ds,k} + \sigma L_r i^{sv}_{dr,k}
\tag{3.45}
$$

As known, $\Psi^{sv}_{ds,k} \approx 0.0$ under SVOC, also, the rotor transient inductance $\left(\sigma L_r\right)$ is very small and can be neglected, so we can consider that:

$$
\Psi^{sv}_{dr,k} = 0 \quad \text{and} \quad \Psi^{sv}_{qr,k} = \left|\overline{\Psi}^{sv}_{r,k}\right|
\tag{3.46}
$$

Assuming that, $K = 1.5p\dfrac{L_m}{\sigma L_s L_r}$, $K' = K\dfrac{L_m}{L_r}\Psi^{sv}_{qr,k+1}$, $K_1 = \dfrac{R_s}{\sigma L_s} + \dfrac{R_r}{\sigma L_r}$,

$K_2 = \dfrac{L_m}{L_r}\omega_{me,k+1}\left(\Psi^{sv}_{qr,k+1}\right)^2$ and $K_3 = \dfrac{R_s L_m}{\sigma L_s L_r}$, equation (3.42) can be reformulated as:

$$
\frac{dT_{d,k+1}}{dt} = -\left(K\frac{L_m}{L_r}\Psi^{sv}_{qr,k+1}\right)u^{sv}_{dr,k} - K_1 T_{d,k+1} - K_2
$$

$$
\frac{dT_{d,k+1}}{dt} = -K'u^{sv}_{dr,k} - K_1 T_{d,k+1} - K_2
$$

$$K'u_{dr,k}^{sv} = -K_1 T_{d,k+1} - \frac{dT_{d,k+1}}{dt} - K_2$$

$$u_{dr,k}^{sv} = \underbrace{-\frac{K_1}{K'} T_{d,k+1} - \frac{1}{K'} \frac{dT_{d,k+1}}{dt}}_{u'_{dr,k+1} = \text{ Active term}} - \underbrace{\frac{K_2}{K'}}_{\Delta u_{dr,k+1}^{sv} = \text{ Compensation term}} \tag{3.47}$$

$$u'_{dr,k+1} = -\frac{K_1}{K'} T_{d,k+1} - \frac{1}{K'} \frac{dT_{d,k+1}}{dt} \tag{3.48}$$

Equation (344) can be reformulated as:

$$\frac{dE_{e,k+1}}{dt} = k\Psi_{qs,k+1}^{sv} u_{qs,k+1}^{sv} - K_1 \Psi_{qs,k+1}^{sv} \Psi_{qs,k+1}^{sv} + K_3 \left(\left(\Psi_{qs,k+1}^{sv} \right)^2 + \left(\Psi_{qr,k+1}^{sv} \right)^2 \right)$$

$$1.2ku_{qr,k+1}^{sv} = -\frac{dE_{e,k+1}}{dt} + 1.44K_1 - 2.88K_3$$

$$u_{qr,k+1}^{sv} = \underbrace{-\frac{1}{1.2k} \frac{dE_{e,k+1}}{dt}}_{u'_{qr,k+1} = \text{ Active term}} + \underbrace{1.2\frac{K_1}{k} - 2.4\frac{K_3}{k}}_{\Delta u_{qr,k+1}^{sv} = \text{ Compensation term}} \tag{3.49}$$

$$u'_{qr,k+1} = -\frac{1}{1.2k} \frac{dE_{e,k+1}}{dt} \tag{3.50}$$

By applying Laplace transform to equations (3.48) and (3.50), assuming the initial torque and air gap energy to be zero.

$$u'_{dr,k+1}(s) = -\frac{K_1}{K'} T_{d,k+1}(s) - \frac{1}{K'} \left[sT_{d,k+1}(s) - T_{d,k+1}(0) \right]$$

$$u'_{dr,k+1}(s) = T_{d,k+1}(s) \left(\frac{-K_1 - s}{K'} \right)$$

$$\frac{T_{d,k+1}(s)}{u'_{dr,k+1}(s)} = \left(\frac{-K_1 - s}{K'} \right) \tag{3.51}$$

In the same manner, we can obtain:

$$\frac{E_{e,k+1}(s)}{u'_{qr,k+1}(s)} = -\frac{1.2k}{s} \tag{3.52}$$

The transfer functions of the of the PI regulators can be expressed as:

$$u'_{dr,k+1}(s) = \overbrace{\left(k_p + \frac{k_i}{s}\right)}^{\text{PI}} \overbrace{\left[T^*_{d,k+1}(s) - T_{d,k+1}(s)\right]}^{d-current\ error} \tag{3.53}$$

$$u'_{qr,k+1}(s) = \underbrace{\left(k_p + \frac{k_i}{s}\right)}_{\text{PI}} \underbrace{\left[E^*_{e,k+1}(s) - E_{e,k+1}(s)\right]}_{q-current\ error} \tag{3.54}$$

Dividing both sides of equations (3.53) and (3.54) by $T_{d,k+1}(s)$ and $E_{e,k+1}(s)$, respectively, it results:

$$\frac{u'_{dr,k+1}(s)}{T_{d,k+1}(s)} = \left(\frac{k_p s + k_i}{s}\right)\left[\frac{T^*_{d,k+1}(s)}{T_{d,k+1}(s)} - 1\right] \tag{3.55}$$

$$\frac{u'_{qr,k+1}(s)}{E_{e,k+1}(s)} = \left(\frac{k_p s + k_i}{s}\right)\left[\frac{E^*_{e,k+1}(s)}{E_{e,k+1}(s)} - 1\right] \tag{3.56}$$

By substituting the term $\{u'_{dr,k+1}(s)\}$ from (3.51) into (3.55), we can obtain:

$$\frac{T_{d,k+1}(s)}{T_{d,k+1}(s)}\left(\frac{s + K_1}{-K'}\right) = \left(\frac{k_p s + k_i}{s}\right)\left[\frac{T^*_{d,k+1}(s)}{T_{d,k+1}(s)} - 1\right]$$

$$\frac{s + K_1}{-K'} = \left(\frac{k_p s + k_i}{s}\right)\left[\frac{T^*_{d,k+1}(s)}{T_{d,k+1}(s)}\right] - \frac{k_p s + k_i}{s}$$

$$\frac{k_p s + k_i}{s}\left[\frac{T^*_{d,k+1}(s)}{T_{d,k+1}(s)}\right] = \frac{k_p s + k_i}{s} - \frac{s + K_1}{K'}$$

$$\frac{T^*_{d,k+1}(s)}{T_{d,k+1}(s)} = \left(\frac{K'k_p s + K'k_i - s^2 - K_1 s}{K's}\right) * \frac{s}{k_p s + k_i}$$

$$\frac{T^*_{d,k+1}(s)}{T_{d,k+1}(s)} = \frac{K'k_p s + K'k_i - s^2 - K_1 s}{K'k_p s + K'k_i}$$

$$\frac{T_{d,k+1}(s)}{T^*_{d,k+1}(s)} = \frac{K'k_p s + K'k_i}{-s^2 + \left(K'k_p - K_1\right)s + K'k_i} \tag{3.57}$$

In the same manner, we get:

$$\frac{E_{e,k+1}(s)}{E^*_{e,k+1}(s)} = \frac{1.2\left(kk_p s + k_i\right)}{-s^2 + 1.2kk_p s + 1.2kk_i} \tag{3.58}$$

The denominator of both (3.57) and (3.58) controls the dynamics of the PI controllers, so it is known as the characteristic equation, its roots must be negative and real
to make the system stable, then the following must be achieved:

$$-s^2 + \left(K'k_p - K_1\right)s + K'k_i = 0 \tag{3.59}$$

By multiplying (3.85) by (-1), we obtain:

$$s^2 + \left(K_1 - K'k_p\right)s - K'k_i = 0 \tag{3.60}$$

For second-order system, the characteristic equation is expressed as:

$$s^2 + 2D\omega_n s + \omega_n^2 = 0 \tag{3.61}$$

Lastly, to evaluate the parameters k_p and k_i of the PI torque regulator, we must perform a comparison between the terms of (3.60) and (3.61), which results in:

$$k_p = \frac{K_1 - 2D\omega_n}{K'} \quad \text{and} \quad k_i = -\frac{\omega_n^2}{K'} \tag{3.62}$$

By performing the same previous steps for the denominator of (3.58), we can find the
parameters k_p and k_i of the PI air-gap energy regulator, as follows:

$$k_p = \frac{-2D\omega_n}{1.2K} \quad \text{and} \quad k_i = -\frac{\omega_n^2}{1.2K} \tag{3.63}$$

Figure 3.6 presents a schematic diagram for the proposed PVC approach, in
which the actual rotor current components can be predicted utilizing equations
(3.25) and (3.26); also, the predicted stator current components can be found
using equations (3.38) and (3.39). Then, the actual values of the torque and the
air-gap energy can be evaluated using (3.40) and (3.43), respectively. As stated
previously, the reference rotor current components are obtained using PI load
power regulator and PI load voltage regulator, respectively. After that, the reference torque can be calculated using (3.34), meanwhile, the reference air-gap
energy can be formulated as:

$$E^*_{e,k+1} = 1.5p\frac{L_m}{\sigma L_s L_r}\left(\Psi^*_{ds,k+1}\Psi^*_{dr,k+1} + \Psi^*_{qs,k+1}\Psi^*_{qr,k+1}\right) \tag{3.64}$$

Now, the torque and gap energy errors are fed to the designed PI torque and air-
gap energy regulators, respectively, to obtain the active voltage components $u'_{dr,k+1}$
and $u'_{qr,k+1}$, and then added to the compensation components $\Delta u^{sv}_{dr,k+1}$ and $\Delta u^{sv}_{qr,k+1}$ to
find the reference rotor voltage components $u^*_{dr,k+1}$ and $u^*_{qr,k+1}$, which are fed with the
actual rotor voltage components to the cost function.

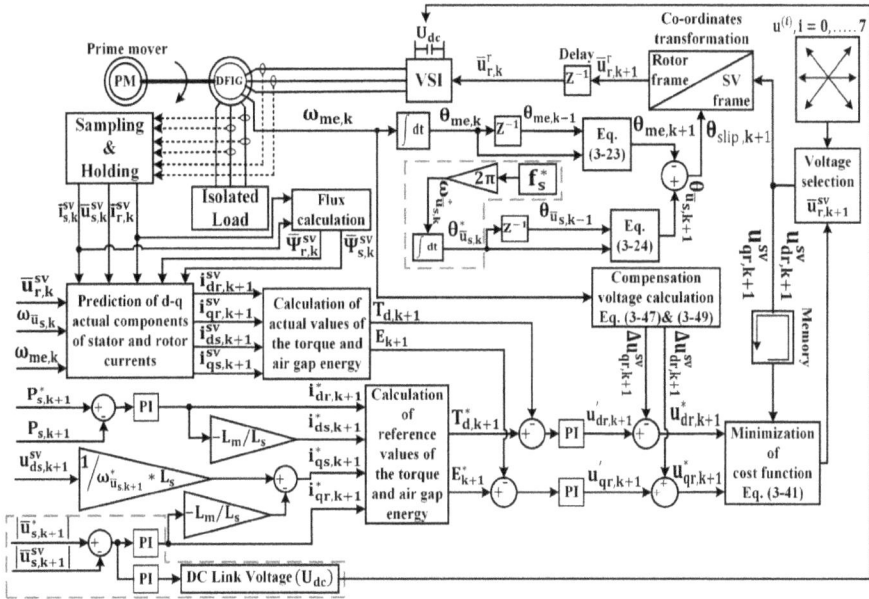

FIGURE 3.6 Scheme of proposed PVC approach for the direct driven standalone DFIG.

3.1.4.2 Proposed PVC Scheme under Wind Driven Operation

The reference components of the rotor voltage $\left(u^*_{dr,k+1} \text{ and } u^*_{qr,k+1}\right)$ can be calculated by the following equations:

$$u^*_{dr,k+1} = R_r i^*_{dr,k+1} + \frac{d\Psi^{sv}_{dr,k+1}}{dt} - \omega_{slip,k+1}\Psi^*_{qr,k+1} \tag{3.65}$$

$$u^*_{qr,k+1} = R_r i^*_{qr,k+1} + \frac{d\Psi^{sv}_{qr,k+1}}{dt} + \omega_{slip,k+1}\Psi^*_{dr,k+1} \tag{3.66}$$

Where the derivative d-q components of the rotor flux are obtained as follows:

$$\frac{d\Psi^{sv}_{dr,k+1}}{dt} = \frac{\Psi^*_{dr,k+1} - \Psi^*_{dr,k}}{T_s} \tag{3.67}$$

$$\frac{d\Psi^{sv}_{qr,k+1}}{dt} = \frac{\Psi^*_{qr,k+1} - \Psi^{sv}_{qr,k}}{T_s} \tag{3.68}$$

The components $\left(\Psi^{sv}_{dr,k} \text{ and } \Psi^{sv}_{qr,k}\right)$ are evaluated using (3.12) and (3.13), meanwhile the calculation of the reference rotor flux components $\left(\Psi^*_{dr,k+1} \text{ and } \Psi^*_{qr,k+1}\right)$ will be described in a systematic manner as follows:

Under SFOC, and steady state operation of the DFIG, we can deduce the following relations:

$$\Psi_{ds,k+1}^{sf} = \left| \overline{\Psi}_{s,k+1}^{sf} \right| \quad \text{and} \quad \Psi_{qs,k+1}^{sf} = 0.0 \tag{3.69}$$

$$u_{ds,k+1}^{sf} \approx 0.0 \quad \text{and} \quad u_{qs,k+1}^{sf} \approx \left| \overline{u}_{s,k+1}^{sf} \right| \tag{3.70}$$

As mentioned in [23], the variation of the rotor flux can be represented as follows:

$$\frac{d\overline{\Psi}_{r,k+1}^{sf}}{dt} = \frac{L_r L_t}{R_r L_m - L_r L_t} \left[\frac{R_r}{L_t} \frac{d\overline{\Psi}_{s,k+1}^{sf}}{dt} - \overline{u}_{r,k+1}^{sf} + j\left(\omega_{\overline{\psi}s,k+1} - \omega_{me,k+1} \right) \overline{\Psi}_{r,k+1}^{sf} \right] \tag{3.71}$$

After that, by performing the Laplace transformation to (3.71), it results in:

$$\overline{\Psi}_{r,k+1}^{sf}(S) = \frac{L_r R_r S \overline{\Psi}_{s,k+1}^{sf}(S) - L_r L_t \overline{u}_{r,k+1}^{sf}(S)}{(R_r L_m - L_r L_t) - jL_r L_t \left(\omega_{\overline{\psi}s,k+1} - \omega_{me,k+1} \right)} \tag{3.72}$$

The time constant of the rotor flux is denoted by $\left(T_f \right)$, and can be defined by:

$$T_f = R_r L_m - L_r L_t \tag{3.73}$$

The magnitude of the stator flux can be obtained as follows:

$$\left| \overline{\Psi}_{s,k+1}^{sf} \right| = \Psi_{ds,k+1}^{sf} = \frac{u_{qs,k+1}^{sf}}{\omega_{\overline{\psi}s,k+1}} = \frac{\left| \overline{u}_{s,k+1}^{sf} \right|}{\omega_{\overline{\psi}s,k+1}}$$

$$= \frac{380}{2 * \pi * 50} = 1.2 \text{ Vs} \tag{3.74}$$

The vectors of the stator and rotor fluxes can be represented in the exponential form as follows:

$$\overline{\Psi}_{s,k+1}^{sf} = \left| \overline{\Psi}_{s,k+1}^{sf} \right| e^{j\omega_{\overline{\psi}s,k+1}t} \tag{3.75}$$

$$\overline{\Psi}_{r,k+1}^{sf} = \left| \overline{\Psi}_{r,k+1}^{sf} \right| e^{j\omega_{me,k+1}t} \tag{3.76}$$

With the aid of (3.75) and (3.76), we can represent the electromagnetic torque of the DFIG by:

$$T_{d,k+1} = 1.5p \frac{L_m}{L_r L_t} \left| \overline{\Psi}_{s,k+1}^{sf} \right| e^{j\omega_{\overline{\psi}s,k+1}t} \times \left| \overline{\Psi}_{r,k+1}^{sf} \right| e^{j\omega_{me,k+1}t} \tag{3.77}$$

where $\times$ refers to the cross product. Consequently, using (3.74) and (3.77), we can express the torque by:

$$T_{d,k+1} = 1.2*1.5p\frac{L_m T_f}{L_r L_t}\left|\overline{\Psi}_{r,k+1}^*\right|\left(1-e^{\frac{-t}{T_f}}\right)*\left(\underbrace{\omega_{\overline{\Psi}s,k+1}-\omega_{me,k+1}}_{\omega_{slip,k+1}}\right) \tag{3.78}$$

The component $\left|\overline{\Psi}_{r,k+1}^*\right|$ can be obtained as follows:

$$\left|\overline{\Psi}_{r,k+1}^*\right| = \frac{L_m}{L_s}\left|\overline{\Psi}_{s,k+1}^*\right|+\sigma L_r\left|\overline{i}_{r,k+1}^*\right| \tag{3.79}$$

where, the component $\left|\overline{\Psi}_{s,k+1}^*\right|$ can be obtained using (3.74).

As it's obvious from (3.78), the DFIG's torque can be controlled via regulating the angular slip frequency $\left(\omega_{slip,k+1}\right)$, with keeping the stator and rotor flux magnitudes constant. Therefore, there will be a reference value of the angular slip speed $\left(\omega_{slip,k+1}^*\right)$, for any value of the reference torque $\left(T_{d,k+1}^*\right)$. After that, the obtained values of $\left(\omega_{slip,k+1}^*\right)$ are used to calculate the values of $\left(\omega_{\overline{\Psi}s,k+1}^*\right)$, which are then utilized to evaluate the rotor flux components $\left(\Psi_{\alpha r,k+1}^* \text{ and } \Psi_{\beta r,k+1}^*\right)$, which are then transferred into synchronous frame and utilized by (3.65) and (3.66) to obtain the d-q reference components of the rotor voltage $\left(u_{dr,k+1}^* \text{ and } u_{qr,k+1}^*\right)$.

The design of the PI torque regulator which will be used to find $\left(\omega_{slip,k+1}^*\right)$ can be performed in the following manner:

As noticed from (3.78), we can consider term $K = 1.2*1.5p\frac{L_m T_f}{L_r L_t}\left|\overline{\Psi}_{r,k+1}^*\right|$ as a constant value, then we differentiate (3.78) with respect to the time, which results:

$$\frac{dT_{d,k+1}}{dt} = K*\omega_{slip,k+1}^**\frac{1}{T_f}e^{\frac{-t}{T_f}} \tag{3.80}$$

After that, by implementing the Laplace transformation to (3.80), it results:

$$ST_{d,k+1}(S)-T_{d,k+1}(0) = \frac{K}{T_f}\frac{1}{S+\frac{1}{T_f}}\omega_{slip,k+1}^*(S) \tag{3.81}$$

With assuming zero initial torque, we get:

$$\omega_{slip,k+1}^*(S) = \frac{T_f}{K}\left(S+\frac{1}{T_f}\right)*ST_{d,k+1}(S) \tag{3.82}$$

$$\frac{T_{d,k+1}(S)}{\omega_{slip,k+1}^*(S)} = \frac{K}{T_f S^2+S} \tag{3.83}$$

We can express the transfer function of the of the PI torque regulator as follows:

$$\omega^*_{slip,k+1}(S) = \overbrace{\left(k_p + \frac{k_i}{S} \right)}^{\text{PI}} \overbrace{\left[T^*_{d,k+1}(S) - T_{d,k+1}(S) \right]}^{\text{Torque error}} \tag{3.84}$$

Dividing both sides of (3.84) by $\{ T_{d,k+1}(S) \}$, it results in:

$$\frac{\omega^*_{slip,k+1}(S)}{T_{d,k+1}(S)} = \left(k_p + \frac{k_i}{S} \right) \left[\frac{T^*_{d,k+1}(S)}{T_{d,k+1}(S)} - 1 \right] \tag{3.85}$$

By substituting the component $\{ \omega^*_{slip,k+1}(S) \}$ from (3.83) into (3.85), we can deduce that:

$$\frac{T_{d,k+1}(S)}{T_{d,k+1}(S)} \left(\frac{T_f S^2 + S}{K} \right) = \left(k_p + \frac{k_i}{S} \right) \left[\frac{T^*_{d,k+1}(S)}{T_{d,k+1}(S)} - 1 \right]$$

$$\frac{T_f S^2 + S}{K} = \left(\frac{k_p S + k_i}{S} \right) \left[\frac{T^*_{d,k+1}(S)}{T_{d,k+1}(S)} \right] - \left(\frac{k_p S + k_i}{S} \right)$$

$$\left(\frac{k_p S + k_i}{S} \right) \left[\frac{T^*_{d,k+1}(S)}{T_{d,k+1}(S)} \right] = \frac{k_p S + k_i}{S} + \frac{T_f S^2 + S}{K}$$

$$\frac{T^*_{d,k+1}(S)}{T_{d,k+1}(S)} = \frac{T_f S^3 + S^2 + K k_p S + K k_i}{K k_p S + K k_i}$$

$$\frac{T_{d,k+1}(S)}{T^*_{d,k+1}(S)} = \frac{K k_p S + K k_i}{T_f S^3 + S^2 + K k_p S + K k_i} \tag{3.86}$$

The denominator of (3.86) represents the characteristic equation, which controls the dynamics of the PI torque regulator, it can be reformulated as:

$$S^3 + \frac{1}{T_f} S^2 + \frac{K k_p}{T_f} S + \frac{K k_i}{T_f} = 0.0 \tag{3.87}$$

For third-order systems, the characteristic equation can be defined by:

$$S^3 + 2D\omega_n S^2 + \omega_n S + \frac{\omega_n^2}{4D^2} = 0 \tag{3.88}$$

Finally, we can easily determine the parameters $(k_p$ and $k_i)$ of the PI torque controller by comparing (3.87) and (3.88), the comparison reveals that:

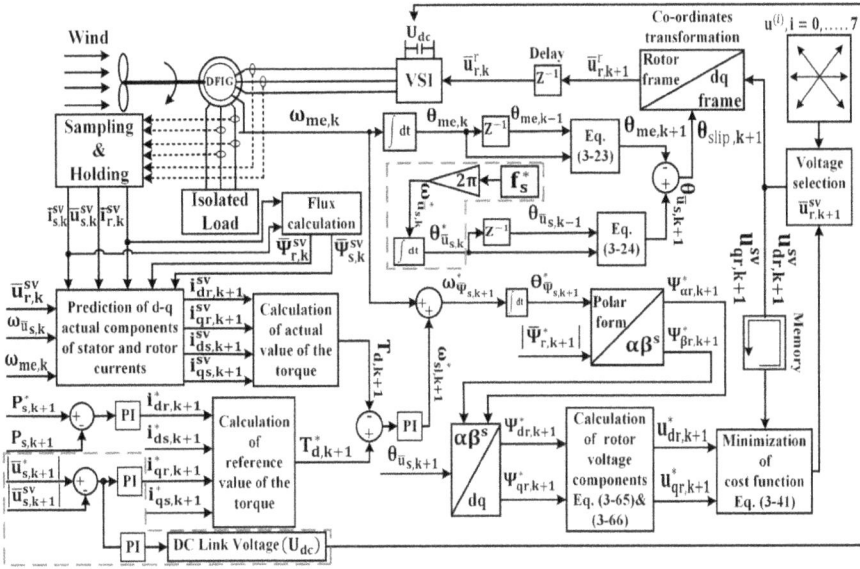

FIGURE 3.7 Scheme of proposed PVC approach for the wind driven standalone DFIG.

$$k_p = \frac{T_f}{K} \omega_n \quad \text{and} \quad k_i = \frac{T_f}{4KD^2} \omega_n^2 \tag{3.89}$$

Now, the PI is used to obtain $\left(\omega_{slip,k+1}^*\right)$, which is then utilized to find $\left(\omega_{\bar\Psi_s,k+1}^*\right)$, which is then integrated to find $\left(\theta_{\bar\Psi_s,k+1}^*\right)$ to be used in calculating the rotor flux components $\left(\Psi_{\alpha r,k+1}^* \text{ and } \Psi_{\beta r,k+1}^*\right)$ as follows:

$$\Psi_{\alpha r,k+1}^* = \left|\bar\Psi_{r,k+1}^*\right| * \cos\left(\theta_{\bar\Psi_s,k+1}^*\right) \tag{3.90}$$

$$\Psi_{\beta r,k+1}^* = \left|\bar\Psi_{r,k+1}^*\right| * \sin\left(\theta_{\bar\Psi_s,k+1}^*\right) \tag{3.91}$$

After that, the rotor flux components are represented in synchronous frame, and then used to obtain the reference voltage components $\left(u_{dr,k+1}^* \text{ and } u_{qr,k+1}^*\right)$, using (3.65) and (3.66).

3.2 ADOPTED CONTROL SCHEME FOR THE LOAD

The cost function which was utilized in this strategy can be defined by:

$$\forall^i = \left|T_{e,k+1}^* - T_{e,k+1}\right|^i + \omega_f \left\|\bar\Psi_{s,k+1}^*\right| - \left|\bar\Psi_{s,k+1}^s\right\|^i \tag{3.92}$$

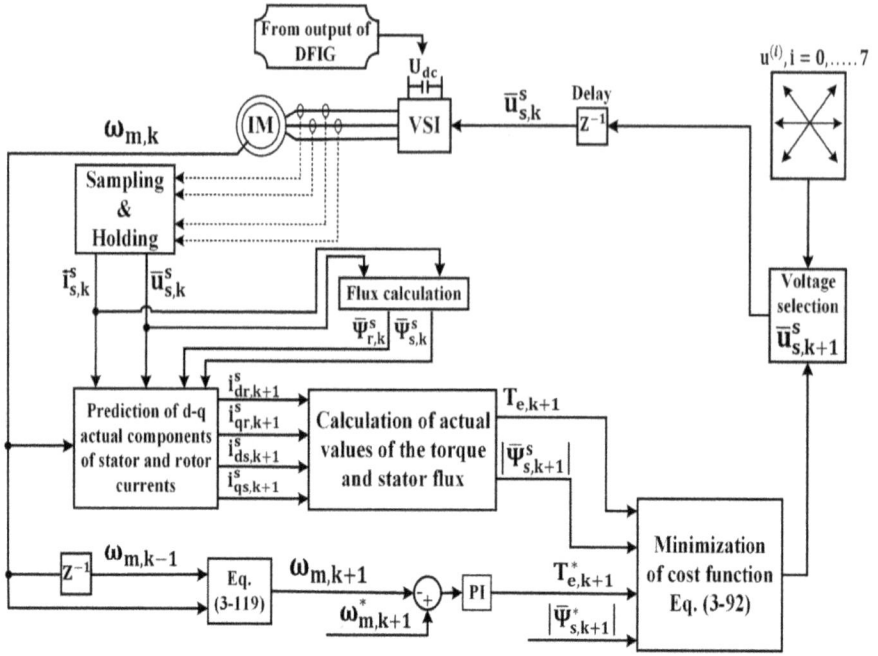

FIGURE 3.8 Scheme of MPDTC algorithm adopted for IM.

As it's obvious in (3–92), the MPDTC depends in its operation on minimizing the error among the reference and predicted components of the torque $\left(T^*_{e,k+1} \text{ and } T_{e,k+1}\right)$, and between the reference and predicted components of the stator flux $\left(\left|\Psi^*_{s,k+1}\right| \text{ and } \left|\Psi^s_{s,k+1}\right|\right)$.

The scheme of MPDTC algorithm is shown in Figure 3.8, in which the stator voltage and stator current are first measured and then sampled. After that, the rotor current can be evaluated in terms of the stator voltage and stator current. The rotor speed $(\omega_{m,k})$ is measured and then can be estimated at instant $(k+1)T_s$ using the following formulation:

$$\omega_{m,k+1} = \omega_{m,k} + \left(\frac{\omega_{m,k} - \omega_{m,k-1}}{\Delta T_s}\right) T_s \tag{3.93}$$

The reference torque $\left(T^*_{e,k+1}\right)$ can be evaluated using a PI speed regulator as shown in the scheme. The prediction of the actual components of the stator current can be implemented as follows:

$$i^s_{ds,k+1} = i^s_{ds,k} + \left(\frac{di^s_{ds,k}}{dt}\right) T_s \tag{3.94}$$

$$i^s_{qs,k+1} = i^s_{qs,k} + \left(\frac{di^s_{qs,k}}{dt}\right) T_s \tag{3.95}$$

The derivatives of the stator current components can be represented via using equation (2.23), as follows:

$$\frac{di^s_{ds,k}}{dt} = \frac{L_{rm}L_{mm}}{L_{sm}L^2_{rm} + L_{rm}L^2_{mm}}\left(\frac{L_{rm}}{L_{mm}}u^s_{ds,k} - \frac{R_{sm}L_{rm}}{L_{mm}}i^s_{ds,k} + R_{rm}i^s_{dr,k} + \omega_{m,k}\Psi^s_{qr,k}\right) \qquad (3.96)$$

$$\frac{di^s_{qs,k}}{dt} = \frac{L_{rm}L_{mm}}{L_{sm}L^2_{rm} + L_{rm}L^2_{mm}}\left(\frac{L_{rm}}{L_{mm}}u^s_{qs,k} - \frac{R_{sm}L_{rm}}{L_{mm}}i^s_{qs,k} + R_{rm}i^s_{qr,k} - \omega_{m,k}\Psi^s_{dr,k}\right) \qquad (3.97)$$

The prediction of the actual components of the rotor current can be implemented using the same previous manner.

The actual torque $\left(T_{e,k+1}\right)$ can be predicted after reformulating equation (2.26) as follows:

$$T_{e,k+1} = 1.5p_{IM}\frac{L_{mm}}{L_{rm}}\left(\Psi^s_{dr,k+1}i^s_{qs,k+1} - \Psi^s_{qr,k+1}i^s_{ds,k+1}\right) \qquad (3.98)$$

The actual value of the stator flux $\left|\Psi^s_{s,k+1}\right|$ can be evaluated as follows:

$$\left|\overline{\Psi}^s_{s,k+1}\right| = \sqrt{\left(\Psi^s_{ds,k+1}\right)^2 + \left(\Psi^s_{qs,k+1}\right)^2} \qquad (3.99)$$

where components $\Psi^s_{ds,k+1}$ and $\Psi^s_{qs,k+1}$ can be found using the prediction values of the stator and rotor currents as follows:

$$\Psi^s_{ds,k+1} = L_s i^s_{ds,k+1} + L_m i^s_{dr,k+1} \qquad (3.100)$$

$$\Psi^s_{qs,k+1} = L_s i^s_{qs,k+1} + L_m i^s_{qr,k+1} \qquad (3.101)$$

3.3 COMPUTATION RESULTS

Tests were performed utilizing MATLAB simulation (Simulink).

3.3.1 EVALUATION THE PERFORMANCE OF THE STANDALONE DFIG UNDER DIRECT-DRIVEN OPERATION

The tests are performed for the four control approaches (SVOC, MPCC, MPDTC, and proposed PVC), through varying the reference active power $\left(P^*_{s,k}\right)$ and driving the DFIG by different operating wind speeds (super-synchronous and sub-synchronous), as shown in Figure 3.9. The reference power and operating speed are varied to evaluate the ability and robustness of each strategy during various operating regimes. The parameters of the DFIG are introduced in Appendix A, in Table A1. The DFIG feeds an isolated load, which is a three-phase induction motor. The parameters of the load and control system are presented in Appendix A, in Tables A2 and A4, respectively.

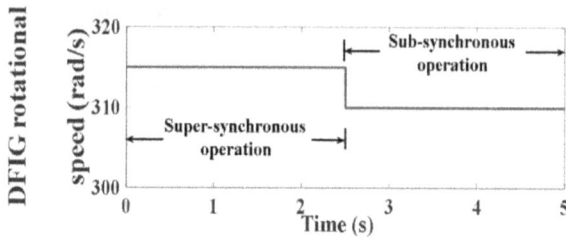

FIGURE 3.9   Prime mover operating speeds (rad/s).

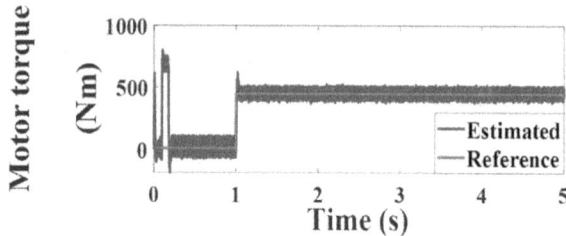

FIGURE 3.10   Active power under SVOC (Watt).

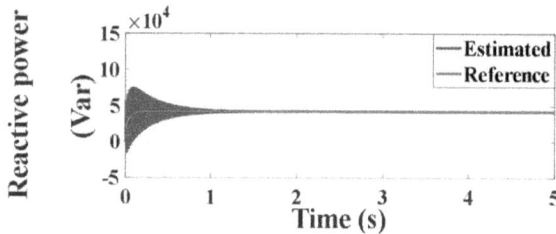

FIGURE 3.11   Reactive power under SVOC (Var).

3.3.1.1   Testing with SVOC Technique

We performed the tests for the DFIG under SVOC principle to study its performance under the stated operating conditions. The obtained results are presented in Figures 3.10–3.19, which show that the actual values follow their reference values smoothly. As known, the VOC is performed by independently controlling the torque current component $\left(i_{dr,k}^{sv}\right)$ and the field current component $\left(i_{qr,k}^{sv}\right)$; which is obvious through Figures 3.16 and 3.17, which confirm that the decoupling between the active and reactive current components has been achieved correctly. Furthermore, the active current component follows the active power and torque changes, while the reactive current component follows the reactive power and rotor flux changes. As noticed from the obtained results, the SVOC is ripple free which is considered as the main merit of the SVOC. However, it suffers from system complexity, also, its dynamic response is slow due to using the PI current regulators.

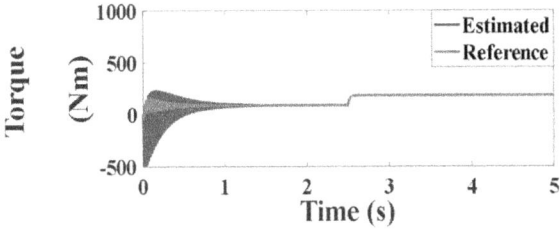

FIGURE 3.12 Torque developed under SVOC (Nm).

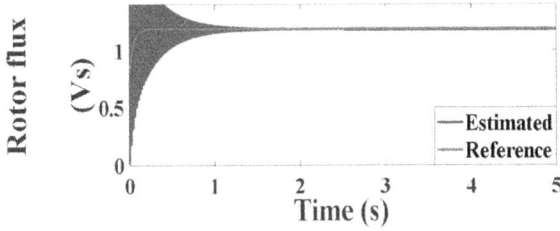

FIGURE 3.13 Rotor flux under SVOC (Vs).

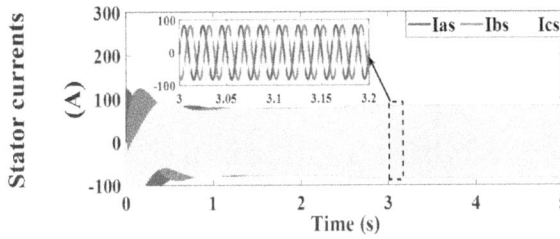

FIGURE 3.14 Stator currents under SVOC (A).

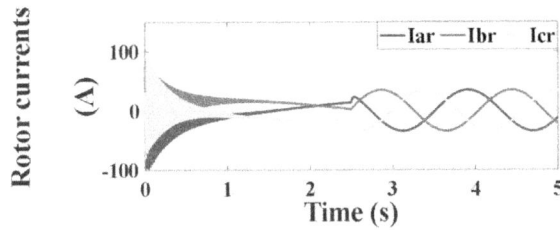

FIGURE 3.15 Rotor currents under SVOC (A).

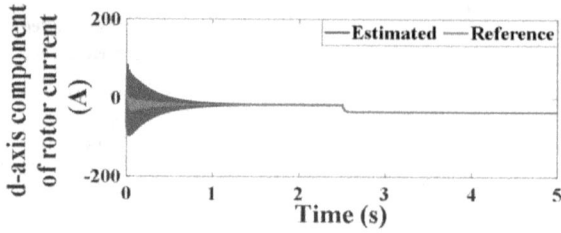

FIGURE 3.16   *d*-axis component of rotor current under SVOC (A).

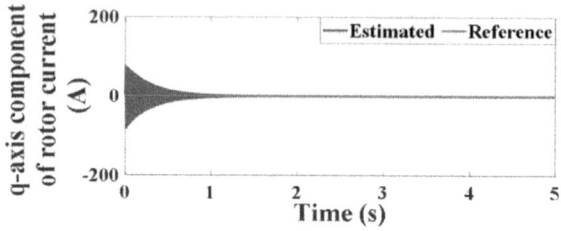

FIGURE 3.17   *q*-axis component of rotor current under SVOC (A).

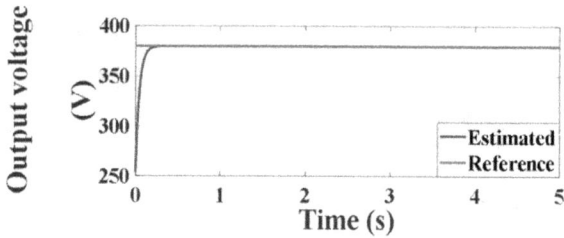

FIGURE 3.18   Load voltage under SVOC (V).

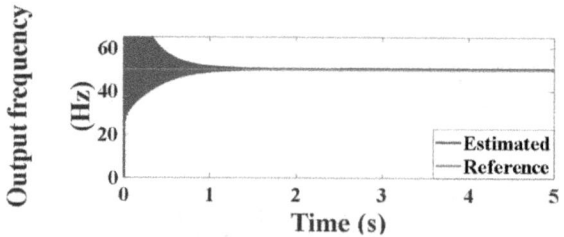

FIGURE 3.19   Load frequency under SVOC (Hz).

3.3.1.2   Testing with MPCC Technique

The DFIG was tested under MPCC approach to study its dynamic performance of the generator and compare it with the formulated predictive controller. The Computation Results are presented in Figures 3.20–3.29, which clarify that the actual values

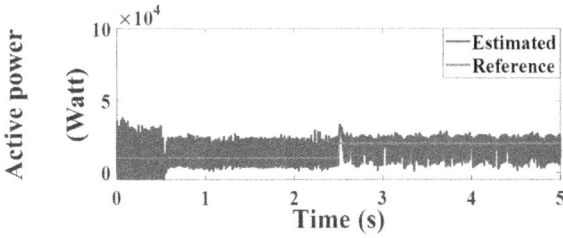

FIGURE 3.20 Active power under MPCC (Watt).

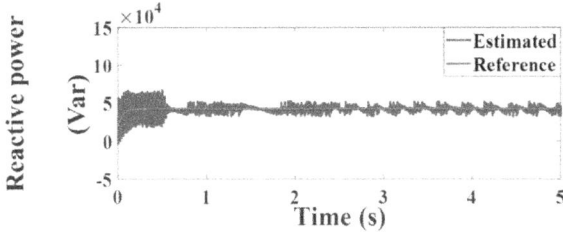

FIGURE 3.21 Reactive power under MPCC (Var).

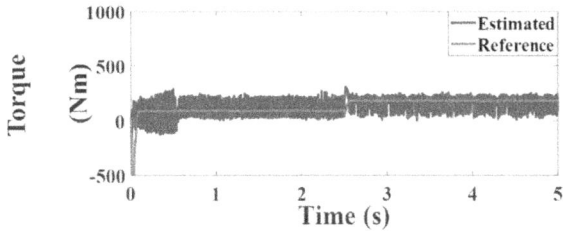

FIGURE 3.22 Torque developed under MPCC (Nm).

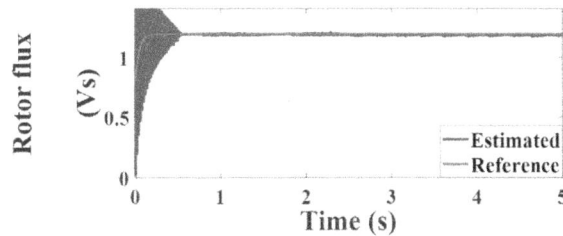

FIGURE 3.23 Rotor flux under MPCC (Vs).

track their reference values with some ripples. Figures 3.26 and 3.27 prove that the decoupling has been correctly accomplished between the current rotor components. Moreover, the torque component $\left(i_{dr,k}^{sv}\right)$ tracks their reference value and also follows the changes of active power and torque, while the field component $\left(i_{qr,k}^{sv}\right)$ tracks their reference value and also follows the changes of reactive power and rotor flux.

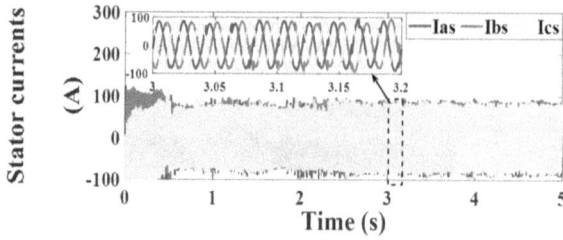

FIGURE 3.24   Stator currents under MPCC (A).

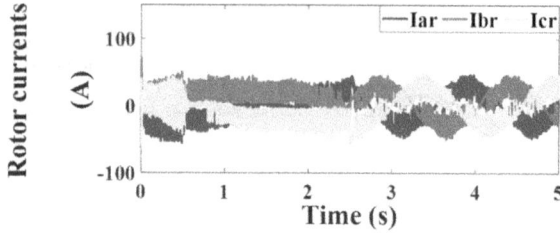

FIGURE 3.25   Rotor currents under MPCC (A).

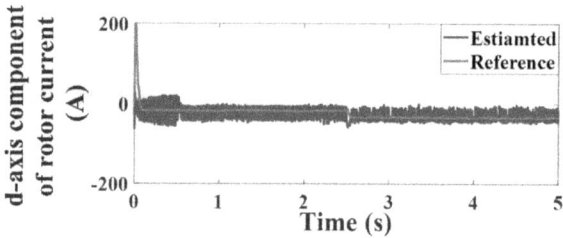

FIGURE 3.26 d-axis component of rotor current under MPCC (A).

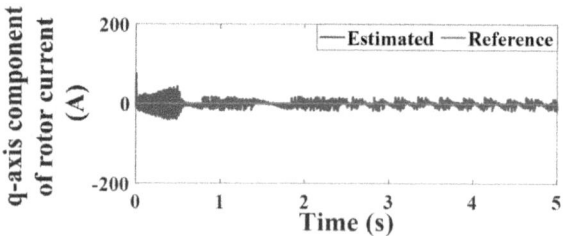

FIGURE 3.27 q-axis component of rotor current under MPCC (A).

3.3.1.3   Testing with MPDTC Technique

Tests for the DFIG's performance were performed under the MPDTC approach and the obtained results are shown in Figures 3.30–3.39, which illustrate that the actual values follow their reference values. Figures 3.36 and 3.37 show that decoupling has

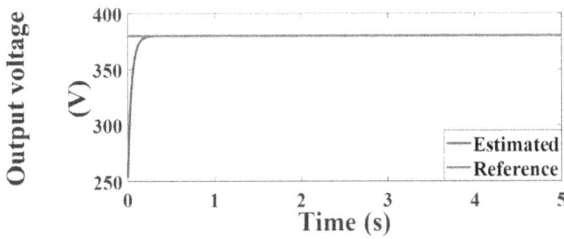

FIGURE 3.28   Load voltage under MPCC (V).

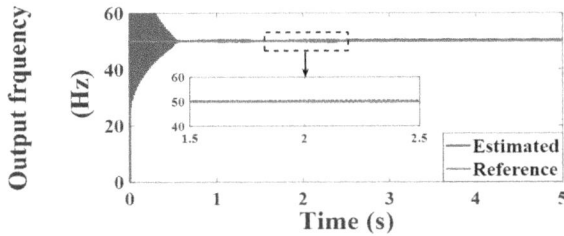

FIGURE 3.29   Load frequency under MPCC (Hz).

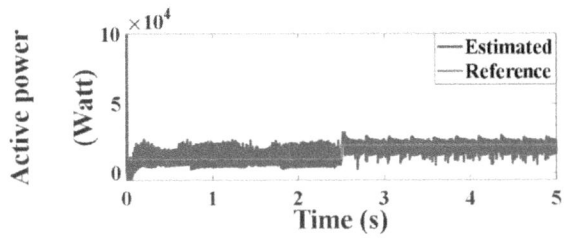

FIGURE 3.30   Active power under MPDTC (Watt).

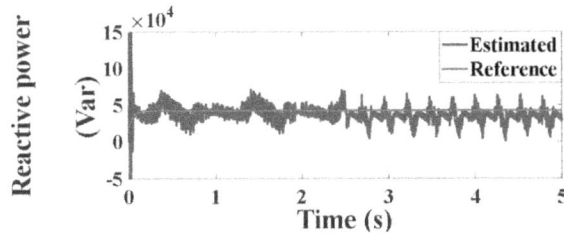

FIGURE 3.31   Reactive power under MPDTC (Var).

been achieved between the current components. Furthermore, the torque component $\left(i_{dr,k}^{sv}\right)$ tracks the active power and torque changes, while the field component $\left(i_{qr,k}^{sv}\right)$ tracks the reactive power and rotor flux changes. The results reveal that the dynamic response of the MPDTC is faster than that of SVOC and MPCC, but it has more ripples compared to SVOC and MPCC.

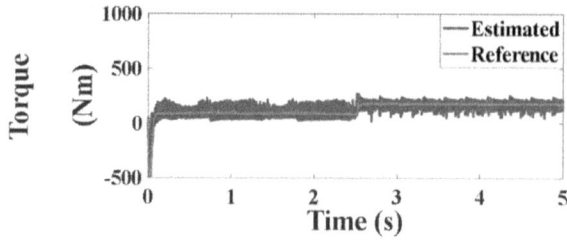

FIGURE 3.32   Torque developed under MPDTC (Nm).

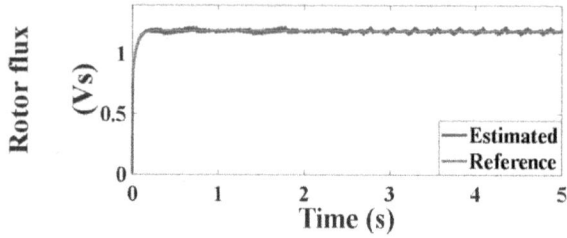

FIGURE 3.33   Rotor flux under MPDTC (Vs).

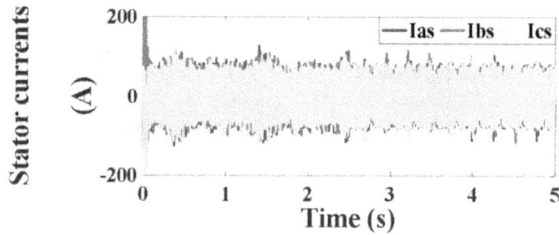

FIGURE 3.34   Stator currents under MPDTC (A).

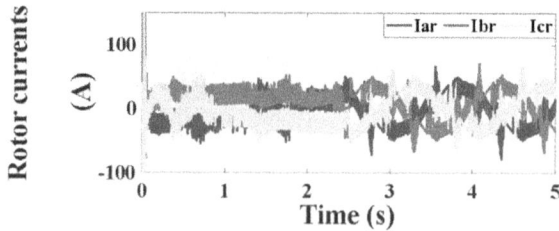

FIGURE 3.35   Rotor currents under MPDTC (A).

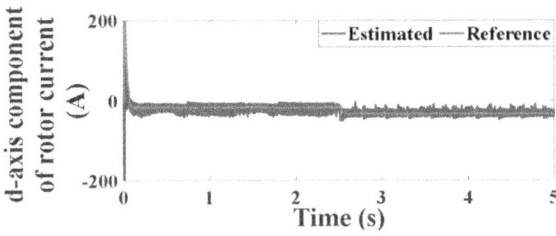

FIGURE 3.36 *d*-axis component of rotor current under MPDTC (A).

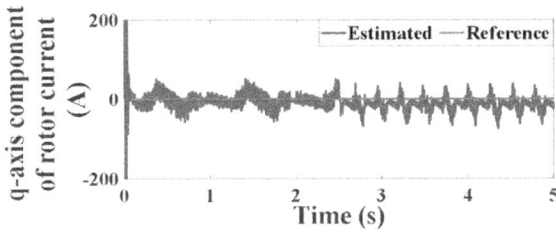

FIGURE 3.37 *q*-axis component of rotor current under MPDTC (A).

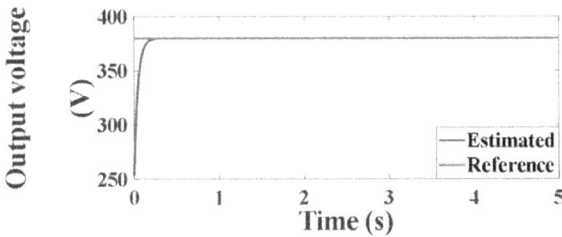

FIGURE 3.38 Load voltage under MPDTC (V).

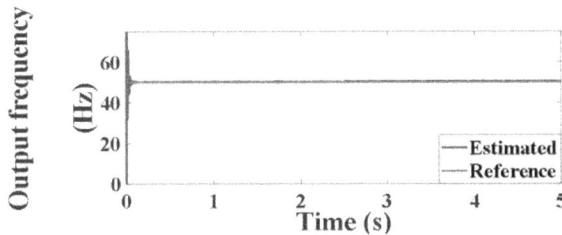

FIGURE 3.39 Load frequency under MPDTC (Hz).

3.3.1.4 Testing with Proposed PVC Technique

The DFIG was tested under the proposed PVC strategy to evaluate its performance. The results are presented in Figures 3.40–3.49, which clarify that the actual values of the active power, reactive power, developed torque, and rotor flux keep track of

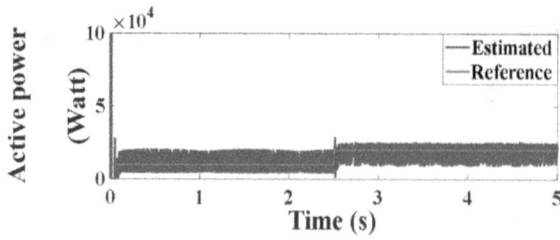

FIGURE 3.40   Active power under PVC (Watt).

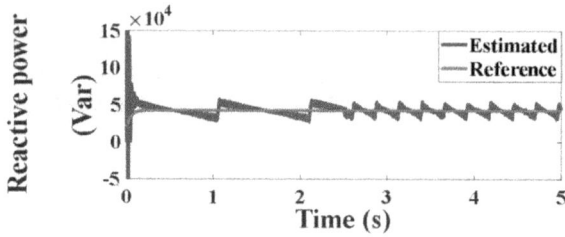

FIGURE 3.41   Reactive power under PVC (Var).

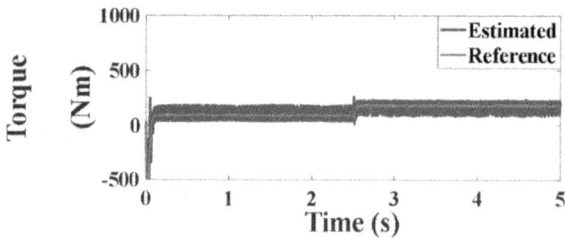

FIGURE 3.42   Torque developed under PVC (Nm).

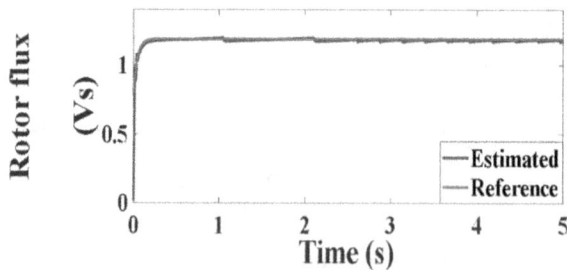

FIGURE 3.43   Rotor flux under PVC (Vs).

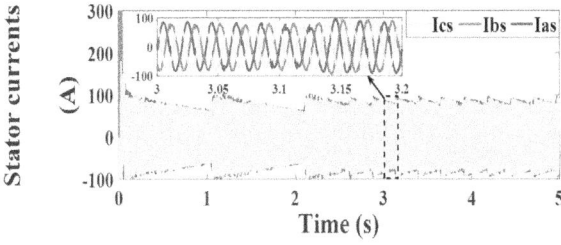

FIGURE 3.44 Stator currents under PVC (A).

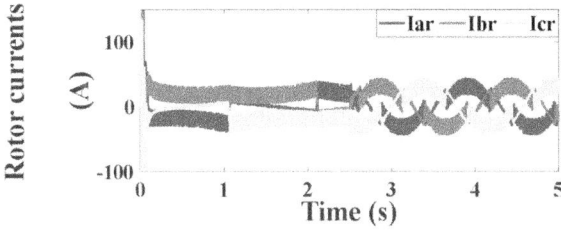

FIGURE 3.45 Rotor currents under PVC (A).

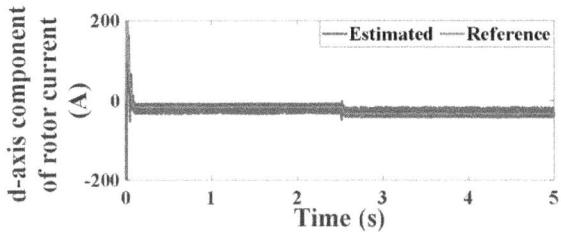

FIGURE 3.46 d-axis component of rotor current under PVC (A).

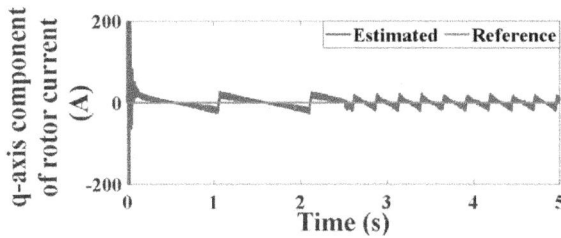

FIGURE 3.47 q-axis component of rotor current under PVC (A).

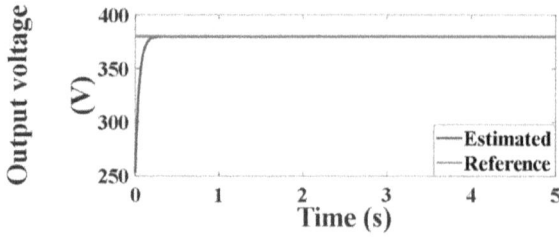

FIGURE 3.48   Load voltage under PVC (V).

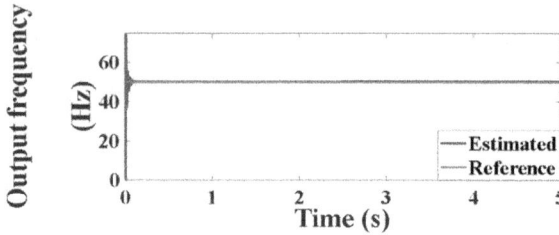

FIGURE 3.49   Load frequency under PVC (Hz).

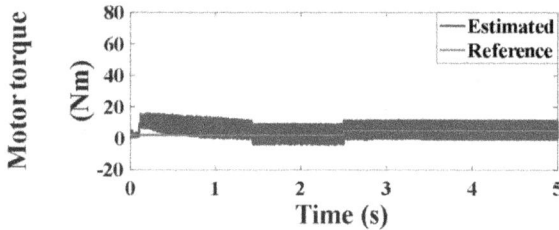

FIGURE 3.50   Motor torque under direct driven operation (Nm).

their reference values. Furthermore, the stator and rotor currents follow the power change, as shown in Figures 3.44 and 3.45. The decoupling has been achieved correctly between the active and reactive current components as shown in Figures 3.46 and 3.47. Moreover, the direct current component follows the active power and torque changes, while the quadrature current component follows the reactive power and rotor flux changes. In addition, Figures 3.48 and 3.49 verify the ability of the control system to ensure a load voltage with constant amplitude and frequency irrespective of the power and speed changes. The results illustrate and prove the effectiveness of the formulated technique, as it has the fastest dynamic response compared to MPDTC, MPCC, and SVOC techniques, moreover, its ripples' content is lower compared with that of MPDTC and MPCC algorithms.

Figures 3.50–3.53, introduce the obtained results which related to the load (IM), which are the motor torque, motor speed, motor stator flux and motor iso stator flux.

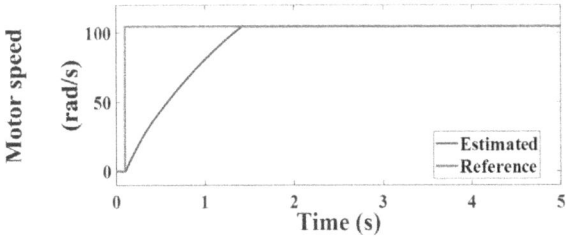

FIGURE 3.51   Motor speed under direct driven operation (rpm).

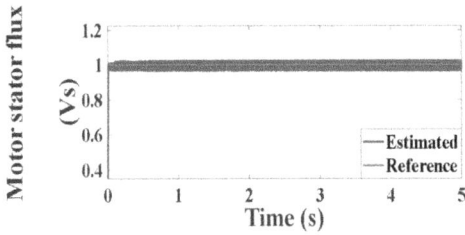

FIGURE 3.52   Motor stator flux under direct driven operation (Vs).

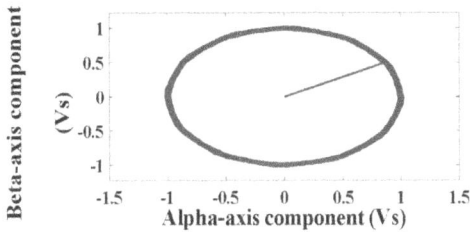

FIGURE 3.53   Motor iso stator flux under direct driven operation (Vs).

3.3.1.5   Comparison Study between All Adopted Control Schemes

Lastly, we performed a comprehensive comparison between the proposed PVC control scheme which was designed in details in our book and the classic control approaches (SVOC, MPCC, and MPDTC) to confirm the validation and effectiveness of our proposed scheme. The techniques' effectiveness was evaluated in terms of dynamic response time, ripples' content, number of executed commutations and THD. The obtained results are shown in Figures 3.54–3.57, which visualize a comprehensive comparison among the four methodologies. Table 3.1 presents a comparison of the dynamic response time for each algorithm to determine which technique takes shorter time to response to the power and torque changes, and therefore easily determine the fastest strategy in dynamic response. The results of Table 3.1 and Figure 3.58 illustrate and confirm that the designed PVC scheme is the fastest dynamic response compared to other adopted controllers, as it takes the shortest response time. Table 3.2 and Figure 3.59, introduces a detailed comparison for the

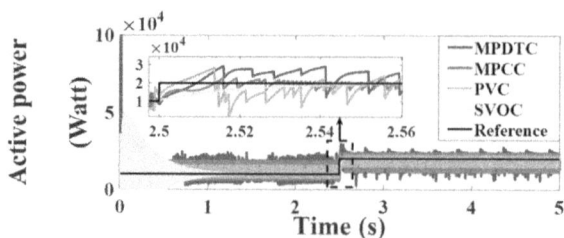

FIGURE 3.54   Active power under direct driven operation (Watt).

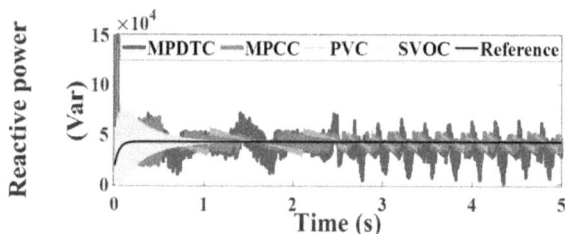

FIGURE 3.55   Reactive power under direct driven operation (Var).

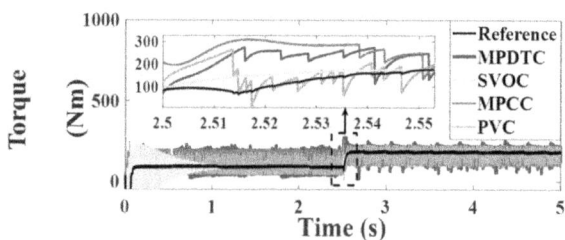

FIGURE 3.56   Developed torque under direct driven operation (Nm).

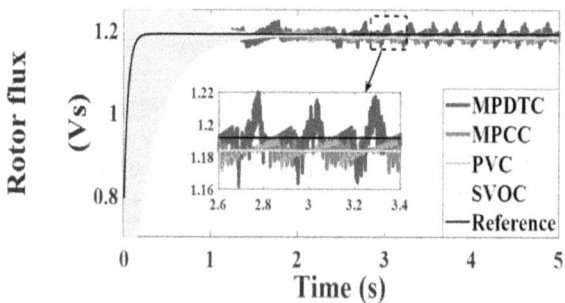

FIGURE 3.57   Rotor flux under direct driven operation (Vs).

TABLE 3.1

Dynamic Response Time Taken by the Actual Values to Track Their References under Direct Driven Operation

Technique	Time Taken by the Active Power Profile (s)	Time Taken by the Torque Profile (s)
SVOC	0.053	0.05
MPCC	0.014	0.049
MPDTC	0.009	0.041
PVC	0.004	0.015

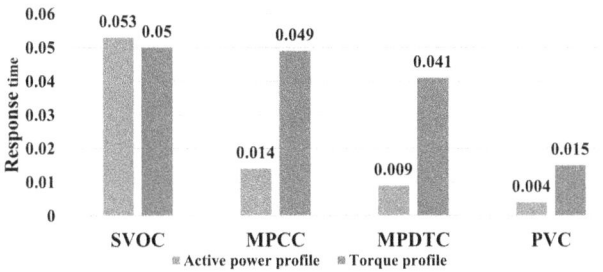

FIGURE 3.58 Histogram of response time taken by the actual values to track their references under direct driven operation (s).

TABLE 3.2

Ripples' Content of the Actual Values above Their References under Direct Driven Operation

Technique	Ripples of Active Power (Watt)	Ripples of Reactive Power (Var)	Ripples of Developed Torque (Nm)	Ripples of Rotor Flux (Vs)
SVOC	860	280	5	0.008
MPCC	8,400	24,450	74.55	0.021
MPDTC	9,470	28,770	90	0.032
PVC	3,950	7,160	40	0.01

four utilized controllers in terms of the ripples' content, and the results reveal that the content of ripples of PVC is lower than that of MPCC and MPDTC algorithms.

The computation time for predictive control schemes is a critical issue, as it is known that model predictive controllers are time-consuming. For this reason, a comparison was also made between the three presented predictive controllers (MPDTC, MPCC, and designed PVC) in terms of number of performed commutations during the code execution. As we know, the switching states of the inverter switches are changed as the voltage index (i) is varied, so we could identify the number of commutations using the voltage index as follows. In the beginning, for each of the three

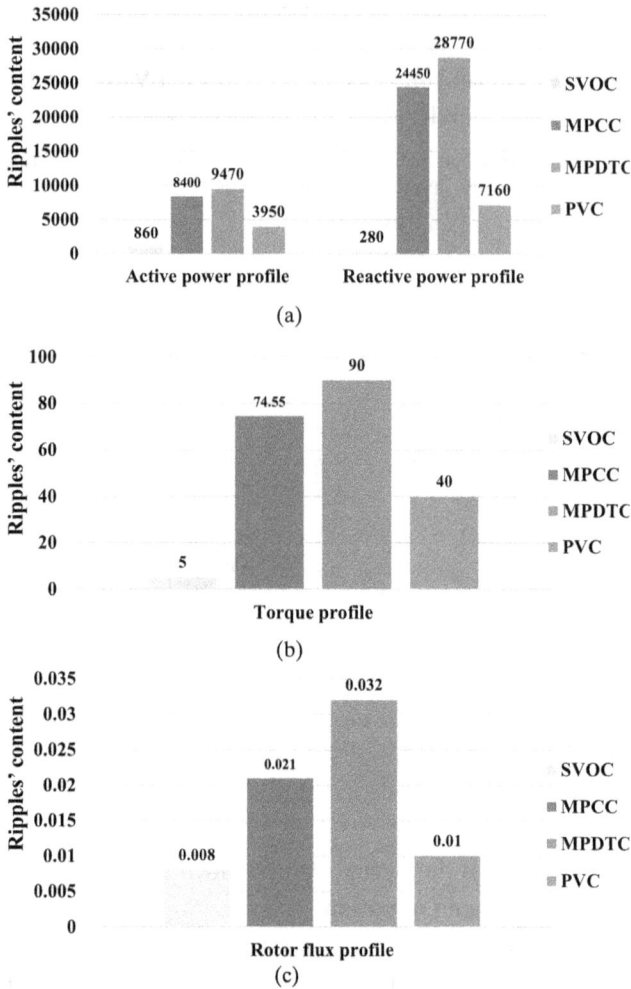

FIGURE 3.59 Histogram of ripples' content of the actual values above their references under direct driven operation. (a) Profiles of active power (Watt) and reactive power (Var). (b) Torque profile (Nm). (c) Rotor flux profile (Vs).

inverter legs (*a*, *b*, and *c*), we calculated the binary variance between each logical state and its previous one; then, we took the summation of the different resultant terms. Finally, we could easily estimate the number of performed commutations in each predictive control technique and performed a comparison between them, as illustrated in Table 3.3 and Figure 3.60. From these analytics, it was revealed that the designed PVC has the lowest number of performed commutations and thus has a shorter computation time compared with the MPDTC and MPCC approaches, which reduces the switching losses of the inverter.

The fast Fourier transform (FFT) analysis for the stator current components under the MPCC, MPDTC and proposed PVC are presented in Figures 3.61–3.69,

TABLE 3.3
Comparison in Terms of Executed Commutations by the Predictive Controllers

Technique	No. of Commutations
MPCC	1,360
MPDTC	1,215
Proposed PVC	513

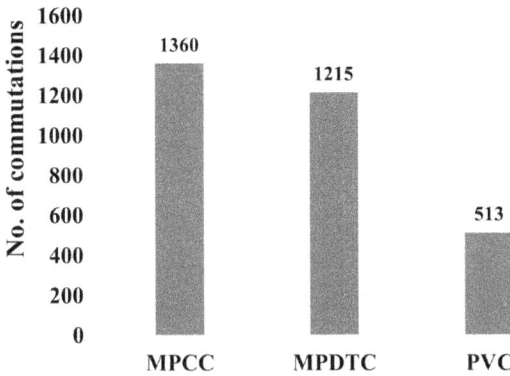

FIGURE 3.60 Histogram of number of commutations executed by the predictive controllers under direct driven operation.

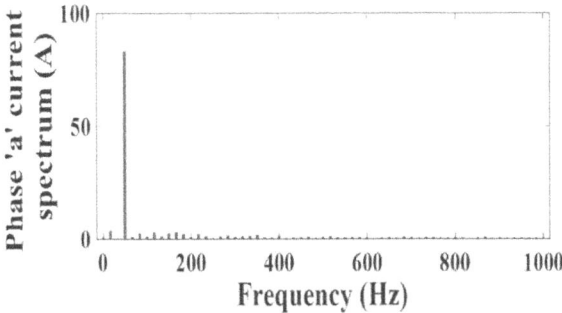

FIGURE 3.61 Spectrum of Phase "a" of stator current under MPCC under direct driven operation.

which clarifies and proves the superiority of the designed PVC over the MPDTC and MPCC, as it has lower THD, which proved also by the numerical values in Table 3.4 and histogram presented in Figure 3.70. Therefore, it can be concluded that the proposed PVC technique is the most convenient approach to be utilized with the DFIG; as it eliminates the system complexity; it is considered as the fastest dynamic

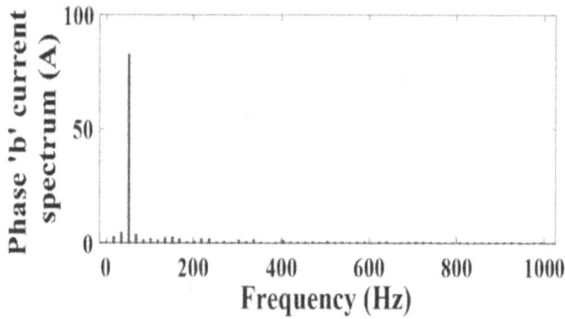

FIGURE 3.62   Spectrum of Phase "b" of stator current under MPCC under direct driven operation.

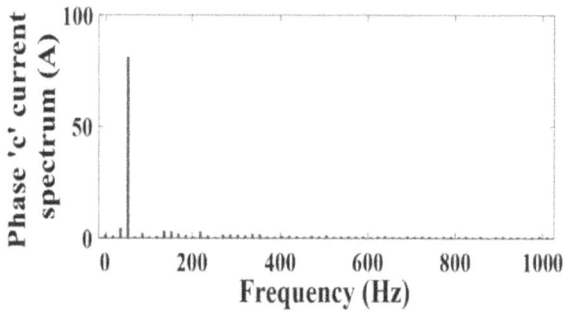

FIGURE 3.63   Spectrum of Phase "c" of stator current under MPCC under direct driven operation.

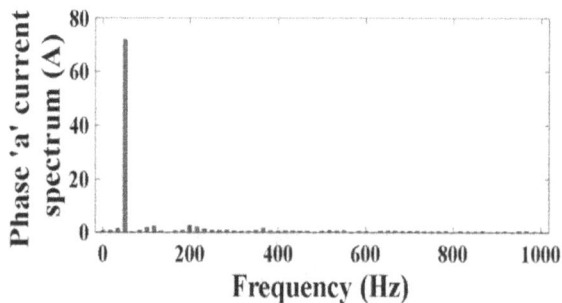

FIGURE 3.64   Spectrum of Phase "a" of stator current under MPDTC under direct driven operation.

response compared to MPDTC, MPCC and SVOC; it has lower ripples' content compared with that of MPDTC and MPCC; it can minimize the computation burden remarkably in terms of reduced number of performed commutations and finally it has lower THD than that of MPCC and MPDTC techniques.

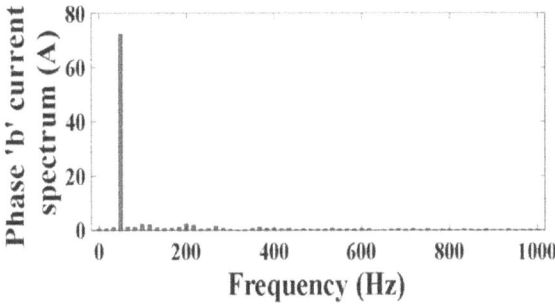

FIGURE 3.65 Spectrum of Phase "b" of stator current under MPDTC under direct driven operation.

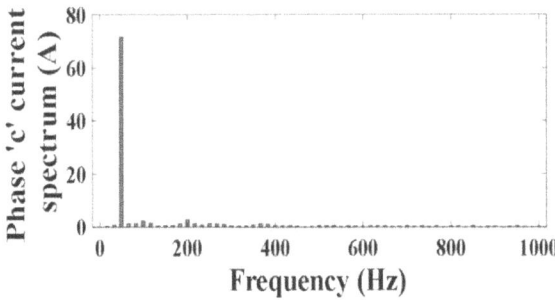

FIGURE 3.66 Spectrum of Phase "c" of stator current under MPDTC under direct driven operation.

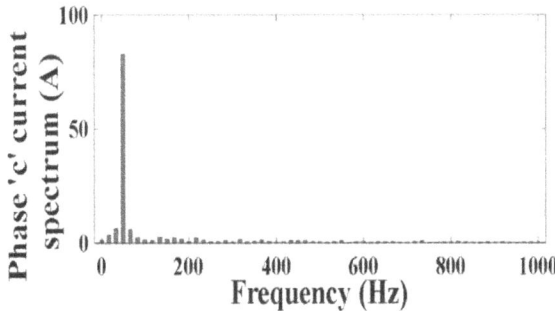

FIGURE 3.67 Spectrum of Phase "a" of stator current under PVC under direct driven operation.

3.3.2 EVALUATION THE PERFORMANCE OF THE STANDALONE DFIG WHICH DRIVEN BY A WIND TURBINE

The dynamic performance of the DFIG was tested under four different control algorithms, which are the SVOC, MPCC, MPDTC methodologies and the proposed PVC technique. The DFIG was driven by a wind turbine with variable speeds, the wind

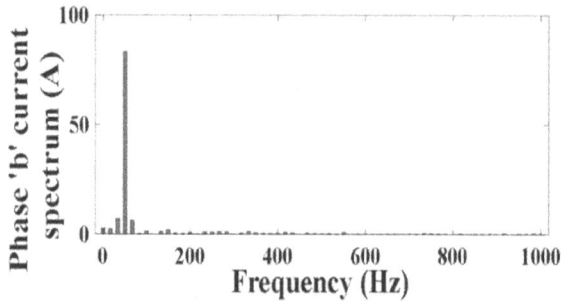

FIGURE 3.68 Spectrum of Phase "b" of stator current under PVC under direct driven operation.

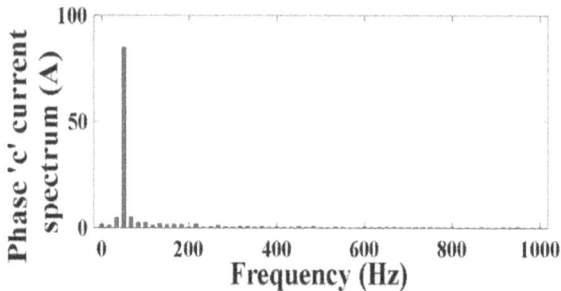

FIGURE 3.69 Spectrum of Phase "c" of stator current under PVC under direct driven operation.

TABLE 3.4
FFT Analysis for Components of the Stator Current

Technique		Phase A	Phase B	Phase C
MPCC	*Fundamental*	83.037 A	82.7778 A	81.144 A
	THD	4.41%	4.90%	4.92%
MPDTC	*Fundamental*	71.9962 A	72.2762 A	71.5451 A
	THD	4.79%	5.09%	5.87%
PVC	*Fundamental*	82.8134 A	83.3904 A	85.0379 A
	THD	3.22%	3.65%	3.72%

speed variation is shown in Figure 3.71, while the operating speeds of the DFIG are shown in Figure 3.72. The generator supplies an isolated load which is a three-phase induction motor. The parameters of the DFIG and wind turbine are presented in Appendix A, in Table A1, meanwhile the data specification of the load and control system are introduced in Appendix A, in Tables A3 and A4, respectively.

3.3.2.1 Testing with SVOC Strategy
Tests were performed for the DFIG under SVOC approach to study its dynamic performance under the mentioned operating conditions. The Computation Results are

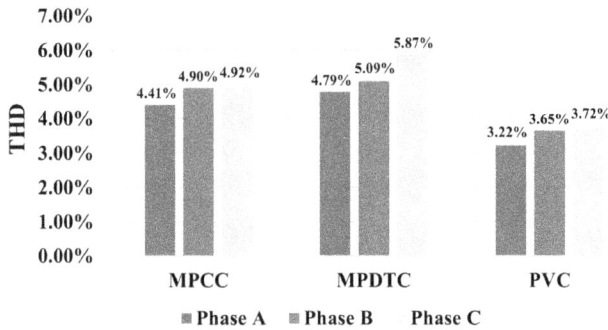

FIGURE 3.70   Histogram of THD analysis for the stator current components under direct driven operation.

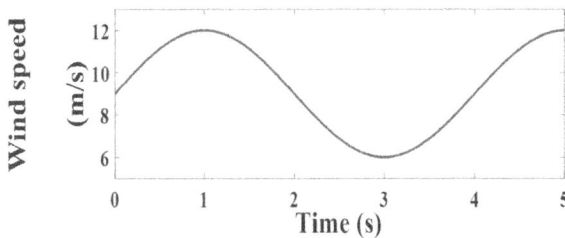

FIGURE 3.71   Wind speed variation (m/s).

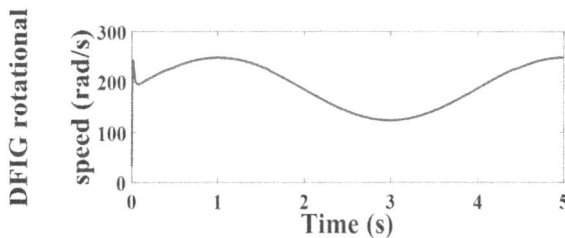

FIGURE 3.72   Operating speeds of the DFIG (rad/s).

shown in Figures 3.73–3.81, which clarify that the actual values track their references in a smooth manner. The obtained results reveal that the SVOC methodology is free of ripples which is considered as the main advantage of this principle. Unfortunately, the system is complex, also, its dynamic response is very slow because of utilizing the PI regulators.

3.3.2.2   Testing with MPCC Strategy

Tests for the DFIG's performance were performed under the MPCC algorithm and the obtained results are shown in Figures 3.82–3.90, which clarify that the actual values follow their reference values. The results show that the dynamic response of the MPCC is faster than that of SVOC, but on the other side, the ripples' content is increased.

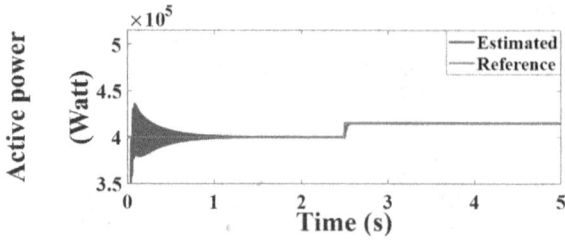

FIGURE 3.73 Active power with SVOC (Watt).

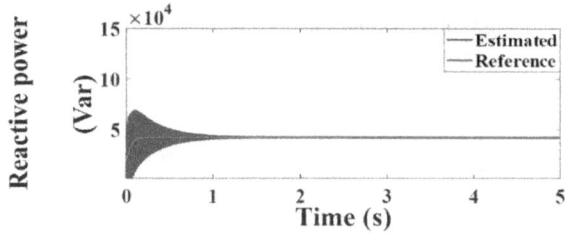

FIGURE 3.74 Reactive power with SVOC (Var).

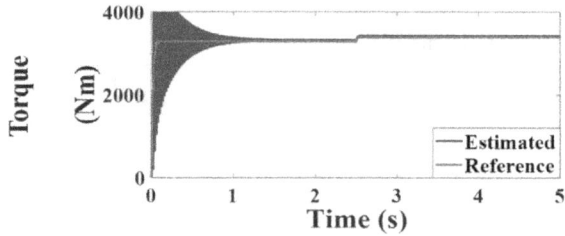

FIGURE 3.75 Developed torque with SVOC (Nm).

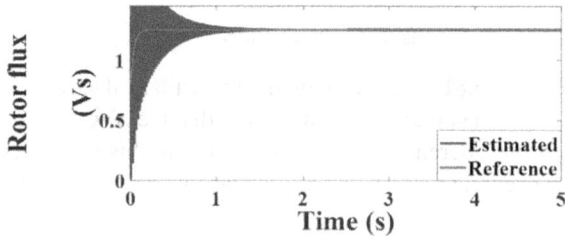

FIGURE 3.76 Rotor flux with SVOC (Vs).

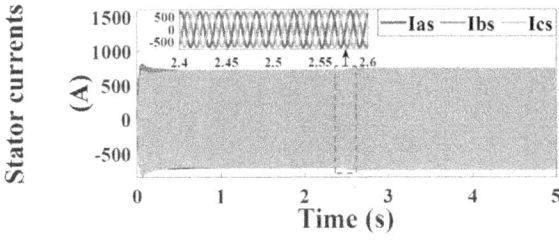

FIGURE 3.77   Stator currents with SVOC (A).

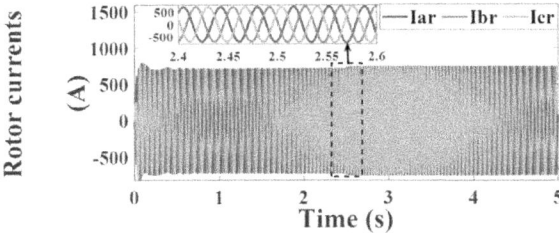

FIGURE 3.78   Rotor currents with SVOC (A).

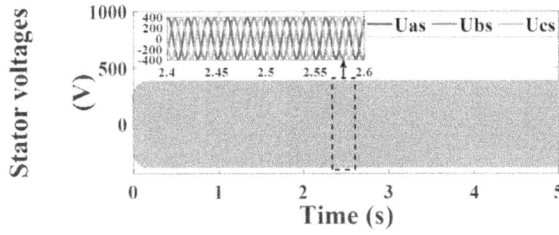

FIGURE 3.79   Stator voltage components with SVOC (V).

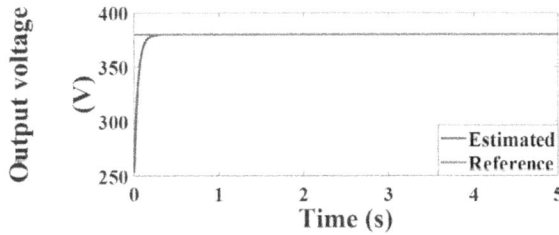

FIGURE 3.80   Load voltage with SVOC (V).

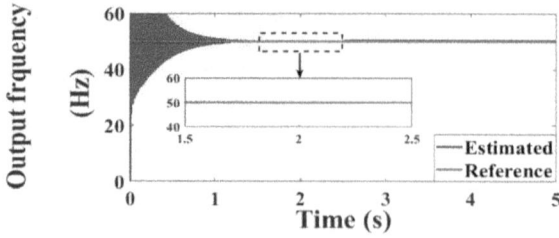

FIGURE 3.81   Frequency of load voltage with MPCC (Hz).

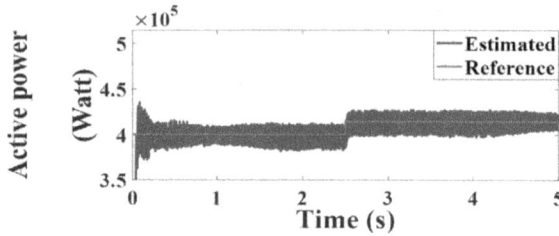

FIGURE 3.82   Active power with MPCC (Watt).

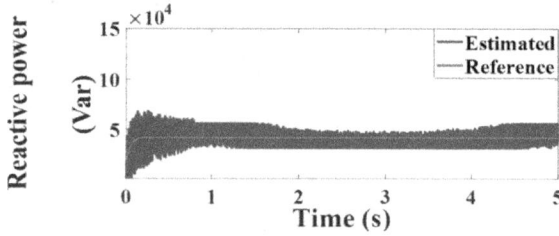

FIGURE 3.83   Reactive power with MPCC (Watt).

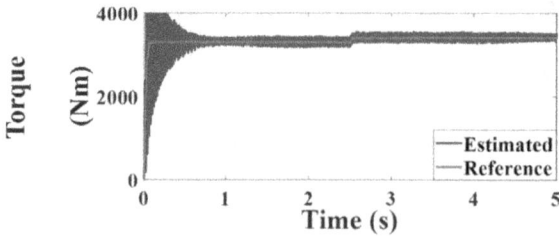

FIGURE 3.84   Developed torque with MPCC (Nm).

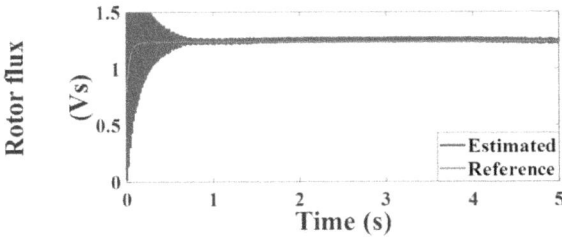

FIGURE 3.85 Rotor flux with MPCC (Vs).

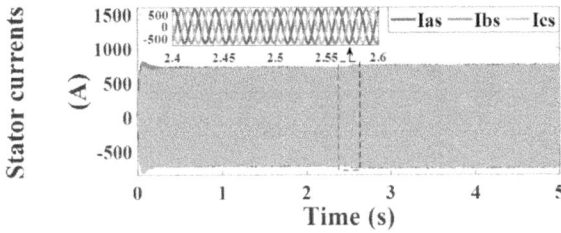

FIGURE 3.86 Stator currents with MPCC (A).

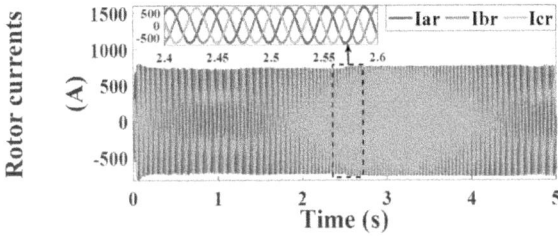

FIGURE 3.87 Rotor currents with MPCC (A).

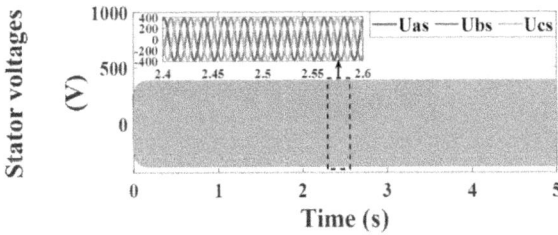

FIGURE 3.88 Stator voltage components with MPCC (V).

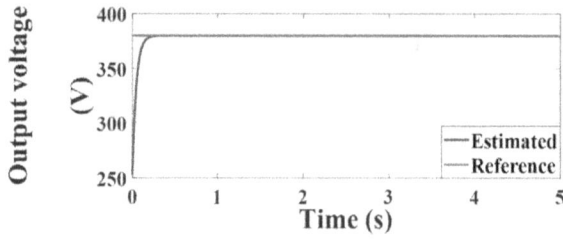

FIGURE 3.89   Load voltage with MPCC (V).

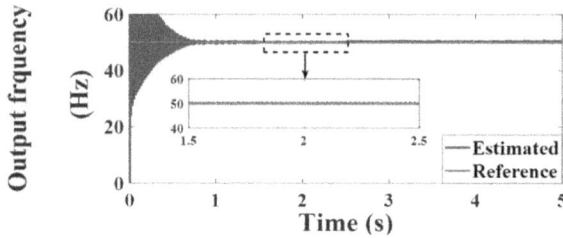

FIGURE 3.90   Frequency of load voltage with MPCC (Hz).

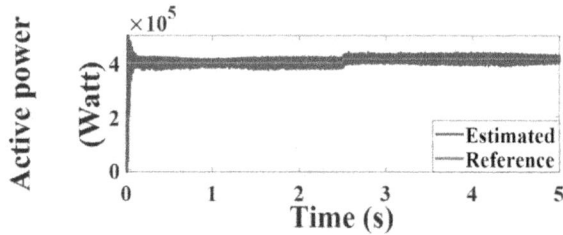

FIGURE 3.91   Active power with MPDTC (Watt).

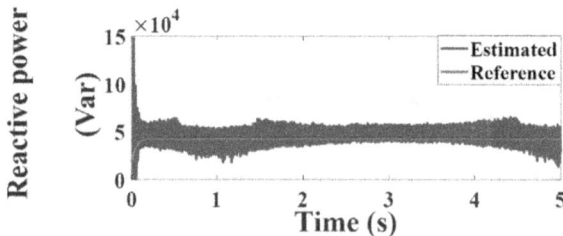

FIGURE 3.92   Reactive power with MPDTC (Var).

3.3.2.3 Testing with MPDTC Strategy

The performance of the DFIG was tested under MPDTC methodology to study the dynamic performance of the generator and compare it with the formulated predictive controller. The obtained results are shown in Figures 3.91–3.99, which show that the

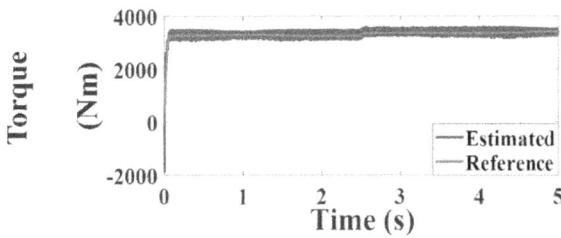

FIGURE 3.93 Developed torque with MPDTC (Nm).

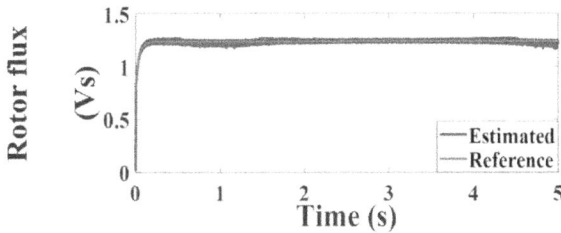

FIGURE 3.94 Rotor flux with MPDTC (Vs).

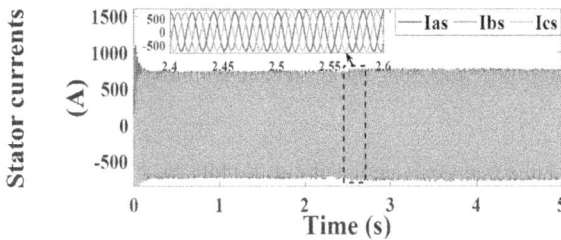

FIGURE 3.95 Stator currents with MPDTC (A).

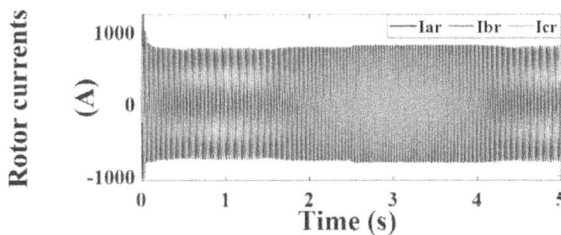

FIGURE 3.96 Rotor currents with MPDTC (A).

actual values follow their reference values with some ripples and with a dynamic response which is faster than that of MPCC and SVOC approaches.

3.3.2.4 Testing with Proposed PVC Strategy

The DFG's dynamic performance was studied under the proposed PVC scheme to prove the ability of the formulated controller to handle the wind changes and

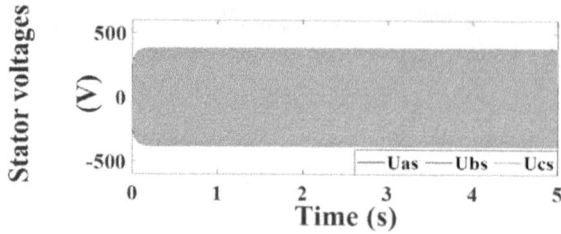

FIGURE 3.97   Stator voltage components with MPDTC (V).

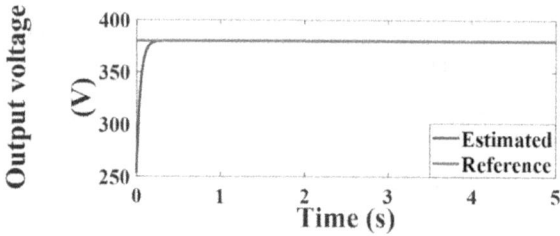

FIGURE 3.98   Load voltage with MPDTC (V).

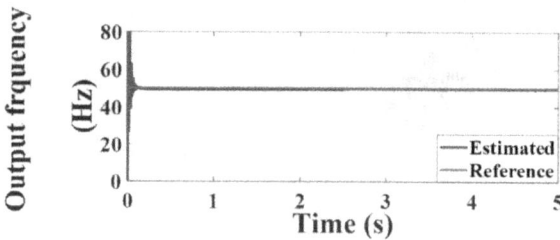

FIGURE 3.99   Frequency of load voltage with MPDTC (Hz).

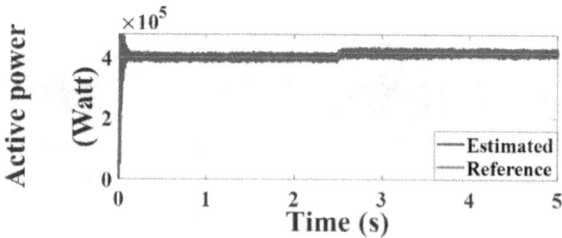

FIGURE 3.100   Active power with PVC (Watt).

enhance the performance of the DFIG. The Computation Results are introduced in Figures 3.100–3.108. The results reveal that the designed controller has succeeded in achieving its targets, as the actual values of the active power, reactive power, developed torque, and rotor flux track their reference values with high precision and lower

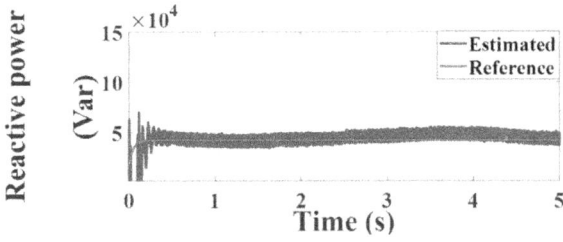

FIGURE 3.101   Reactive power with PVC (Var).

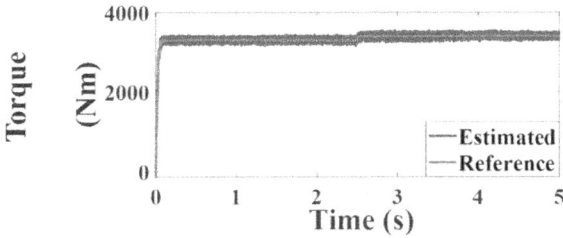

FIGURE 3.102   Developed torque with PVC (Nm).

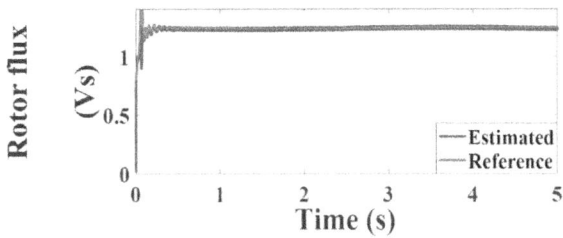

FIGURE 3.103   Rotor flux with PVC (Vs).

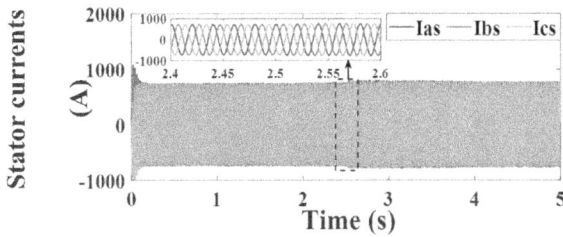

FIGURE 3.104   Stator currents with PVC (A).

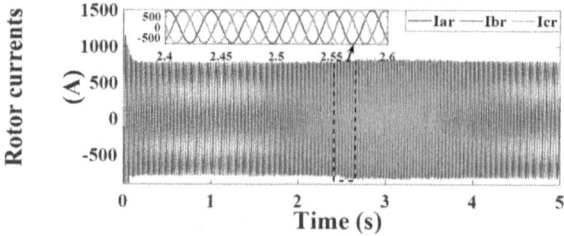

FIGURE 3.105 Rotor currents with PVC (A).

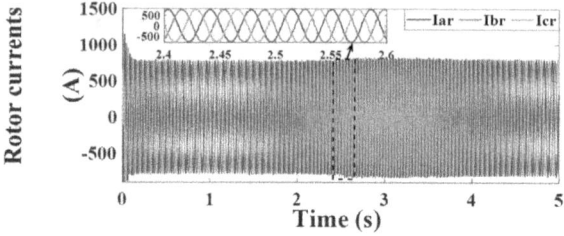

FIGURE 3.106 Stator voltage components with PVC (V).

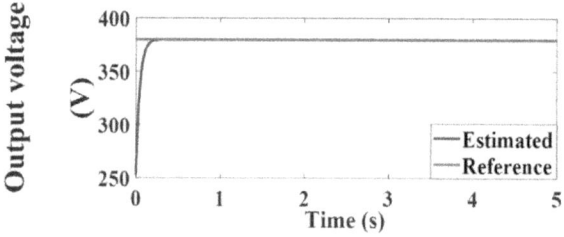

FIGURE 3.107 Load voltage with PVC (V).

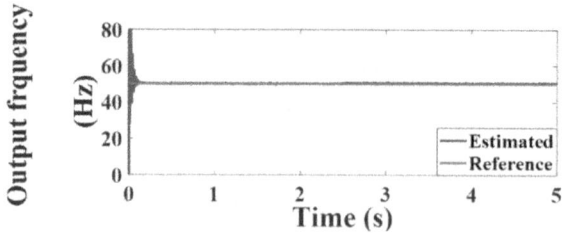

FIGURE 3.108 Frequency of load voltage with PVC (Hz).

ripples' content compared with MPDTC and MPCC. Furthermore, the DFIG under the designed scheme could provide a constant output voltage with constant frequency even under wind changes, which is considered as a principal requirement for stand-alone systems. As it's obvious from the obtained results, the formulated controller has successfully improved the dynamic response of DFIG and made it faster than that of SVOC, MPCC, and MPDTC techniques.

Figures 3.109–3.112, present the results which are related to the load (IM), which are the motor torque, motor speed, motor stator flux, and motor iso stator flux.

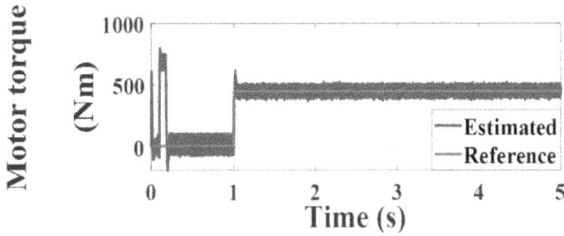

FIGURE 3.109   Motor torque (Nm).

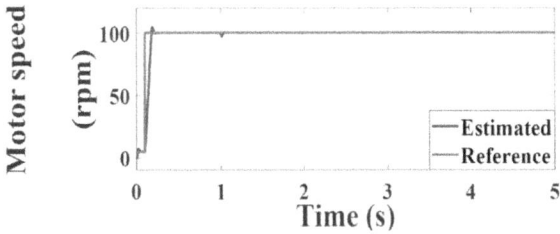

FIGURE 3.110   Motor speed (rpm).

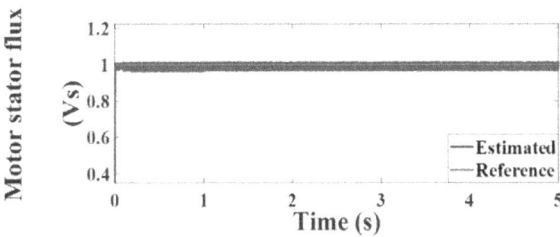

FIGURE 3.111   Motor stator flux (Vs).

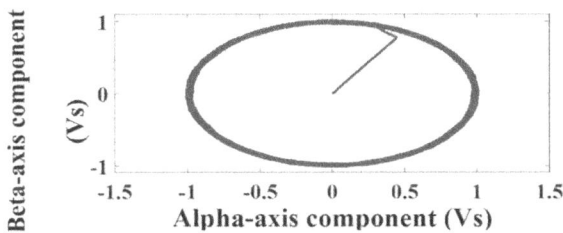

FIGURE 3.112   Motor iso stator flux (Vs).

3.3.2.5 Comparison Study

After testing and studying the performance of the DFIG under each algorithm individually, we preferred to perform a comprehensive and detailed comparison for the
DFIG's performance under SVOC, MPDTC, MPCC, and our formulated controller,
so we can easily define the advantages and shortages of each control scheme to be
able to easily determine the most appropriate methodology to be utilized with the
DFIG. The comparison has been performed in terms of time of dynamic response,
content of ripples, number of executed commutations, and THD. The results of the
comparison are presented in Figures 3.113–3.116. The obtained results reveal that
the proposed PVC algorithm has a faster dynamic response than that of MPDTC,
MPCC, and SVOC, as it takes less time to reach the required reference value, which
is also evident in Table 3.5 and Figure 3.117. Furthermore, Table 3.6 clarifies that
our designed controller can minimize the content of ripples remarkably compared
with MPDTC and MPCC methods, which is also confirmed via histogram presented
in Figure 3.118. Moreover, the proposed algorithm introduces less computational

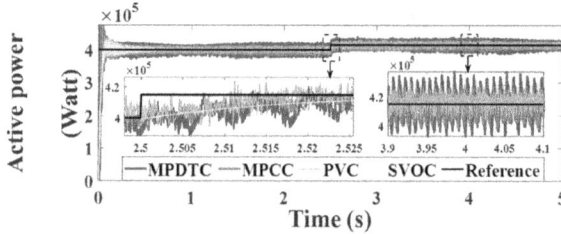

FIGURE 3.113   Active power (Watt).

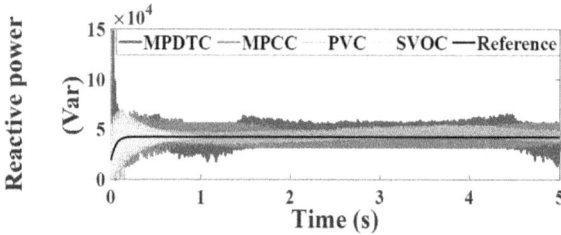

FIGURE 3.114   Reactive power (Var).

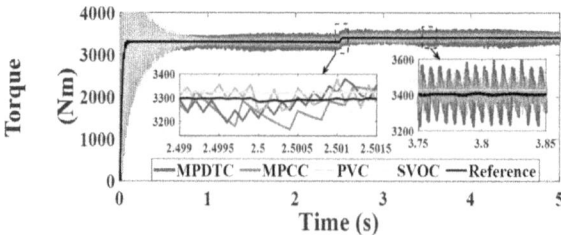

FIGURE 3.115   Developed torque (Nm).

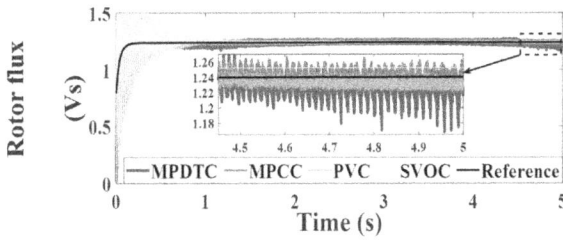

FIGURE 3.116 Rotor flux (Vs).

TABLE 3.5

Dynamic Response Time Taken by the Actual Values to Track Their References

Technique	Time Taken by the Active Power Profile (ms)	Time Taken by the Torque Profile (ms)
SVOC	55	11
MPCC	20	1.2
MPDTC	15	0.5
PVC	11	0.2

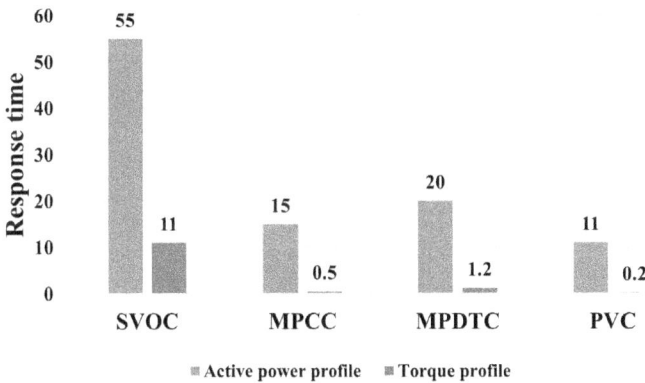

FIGURE 3.117 Histogram of response time taken by the actual values to track their references (s).

TABLE 3.6

Ripples' Content of the Actual Values above Their References

Algorithm	Ripples of Active Power (Watt)	Ripples of Reactive Power (Var)	Ripples of Developed Torque (Nm)	Ripples of Rotor Flux (Vs)
SVOC	1,600	1,220	24	0.005
MPCC	15,600	13,820	108	0.02
MPDTC	22,300	20,350	189	0.031
PVC	9,800	5,610	61	0.009

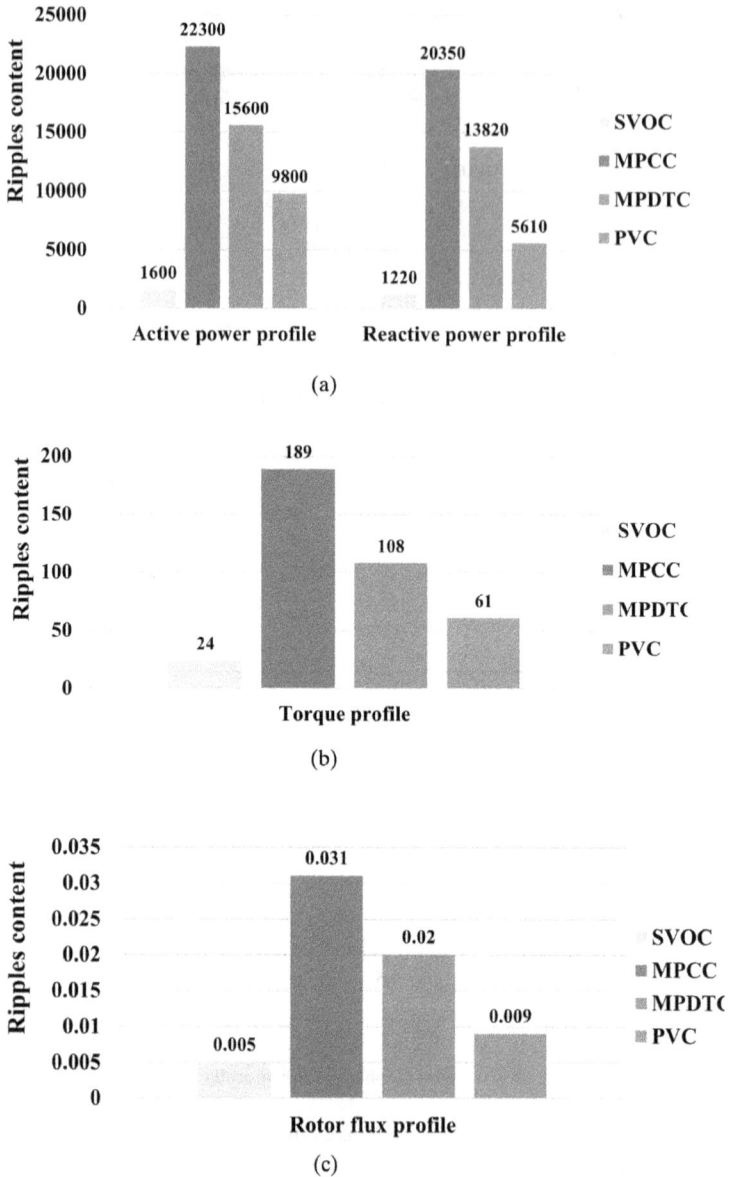

FIGURE 3.118 Histogram of ripples' content of the actual values above their references. (a) Profiles of active power (Watt) and reactive power (Var). (b) Torque profile (Nm). (c) Rotor flux profile (Vs).

burden, as it has a lower number of executed commutations compared with that of MPDTC and MPCC strategies, as shown in Table 3.7 and Figure 3.119.

The FFT analysis for the components of the stator current under the MPCC, MPDTC, and our formulated PVC are presented in Figures 3.120–3.128. The THD of the designed PVC is lower than that of MPDTC and MPCC as it's clear from

TABLE 3.7
Comparison in Terms of Commutations Executed by the Predictive Controllers

Technique	No. of Commutations
MPCC	4,087
MPDTC	3,856
Proposed PVC	2,412

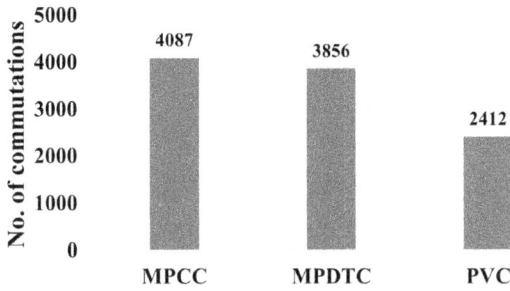

FIGURE 3.119 Histogram of number of commutations executed by the predictive controllers.

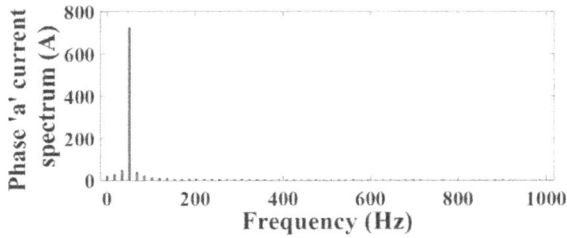

FIGURE 3.120 Spectrum of Phase "a" of stator current under MPCC.

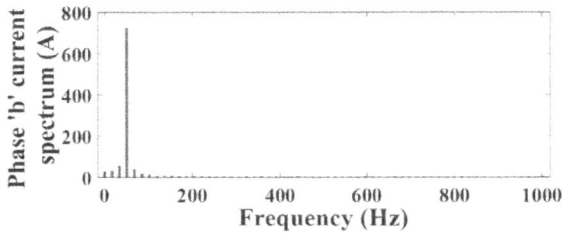

FIGURE 3.121 Spectrum of Phase "b" of stator current under MPCC.

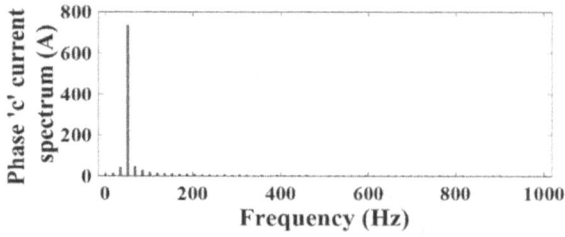

FIGURE 3.122   Spectrum of Phase "c" of stator current under MPCC.

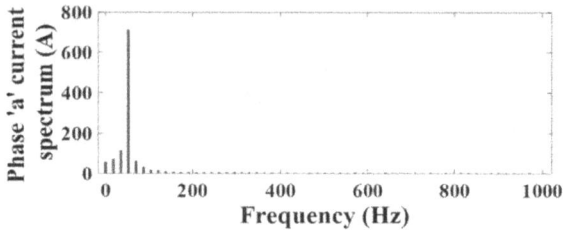

FIGURE 3.123   Spectrum of Phase "a" of stator current under MPDTC.

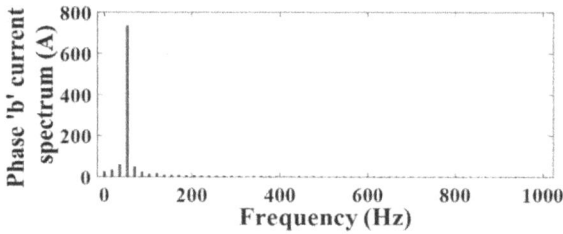

FIGURE 3.124   Spectrum of Phase "b" of stator current under MPDTC.

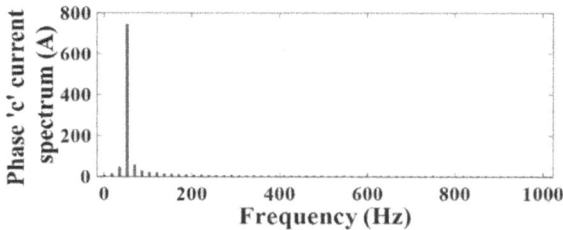

FIGURE 3.125   Spectrum of Phase "c" of stator current under MPDTC.

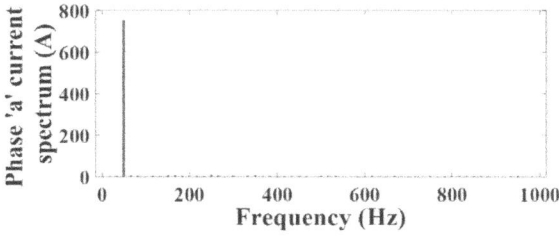

FIGURE 3.126   Spectrum of Phase "a" of stator current under PVC.

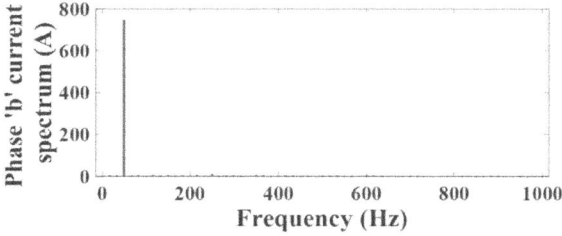

FIGURE 3.127   Spectrum of Phase "b" of stator current under PVC.

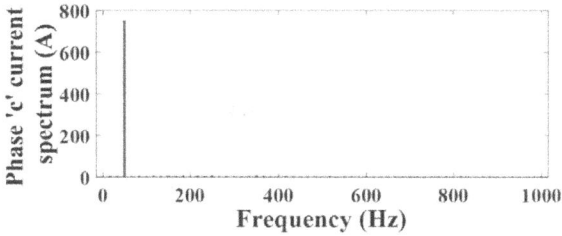

FIGURE 3.128   Spectrum of Phase "c" of stator current under PVC.

comparing the current spectrums and also confirmed through the numerical values which are presented in Table 3.8 and Figure 3.129.

Eventually, it can be deduced that, the proposed PVC scheme is the best selection for using with the DFIG, as it obviated the main defects which faced other classic controllers and succeeded in improving the dynamic performance of the generator, as explained in details.

As conclusion, we can brief the main differences between the adopted predictive control techniques in the following Table 3.9, which clarifies the superiority of our proposed PVC scheme.

TABLE 3.8

FFT Analysis for the Stator Current Components

		Phase A	Phase B	Phase C
MPCC	*Fundamental*	725.288 A	724.037 A	734.196 A
	THD	2.78%	2.15%	3.85%
MPDTC	*Fundamental*	712.155 A	736.363 A	745.412 A
	THD	3.02%	2.92%	4.11%
PVC	*Fundamental*	751.477 A	746.606 A	750.4 A
	THD	1.08%	1.06%	1.23%

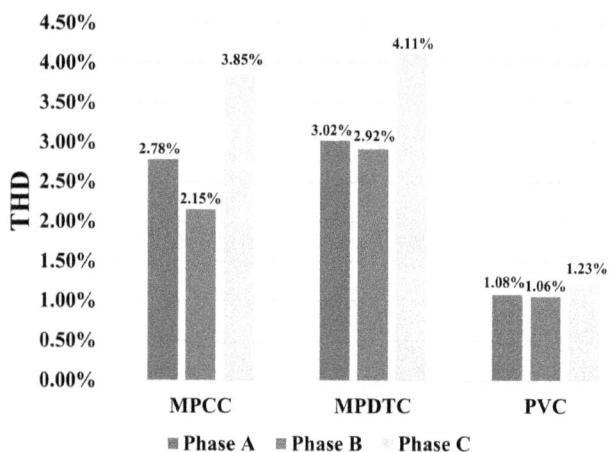

FIGURE 3.129   Histogram of THD analysis for the stator current components.

TABLE 3.9

Comparison between the Adopted Predictive Control Techniques

Technique	Simplicity	Dynamic Response	Ripples Content	Computational Burden	THD
MPCC	It suffers from system complexity, as it has estimated variables	Faster than SVOC but slower than MPDTC& PVC	Lower than MPDTC but higher than SVOC& PVC	The highest	Lower than MPDTC but higher than PVC
MPDTC	It suffers from system complexity, as it has estimated variables and weighting factor (ω_f)	Faster than SVOC and MPCC but slower than MPDTC& PVC	The highest	Lower than MPCC but higher than PVC	The highest
Proposed PVC	(The simplest technique) - No weighting factor (ω_f) - No estimated variables - It uses FCS	The fastest	The least	The least	The least PCC

4 Dynamic Performance Analysis of an Alternative Operating Regime of DFIG (Grid-Connected Case)

4.1 ADOPTED CONTROL SCHEMES FOR THE GRID-CONNECTED DFIG

4.1.1 SVOC APPROACH

The operating principle of the SVOC algorithm was explained in a systematic manner in the previous chapter, in Section 3.2.1. There are some differences between grid-connected operation and standalone operation of the DFIG: as mentioned previously, in standalone systems, the d-q reference rotor current components were found using the errors of the load active power and load voltage with the aid of two PI regulators to provide the required power of the load and maintain a constant voltage for the load, moreover, the reference load frequency is used to calculate the reference synchronous angular speed to keep the frequency of the load voltage constant; however, in grid-connected systems, the voltage and frequency of the grid are already constant, so the reference components of the rotor current $\left(i_{dr,k}^{*} \text{ and } i_{qr,k}^{*}\right)$ can be obtained directly in terms of reference active power $\left(P_{s,k}^{*}\right)$ and reference reactive power $\left(Q_{s,k}^{*}\right)$, with the help of equations (3.9) and (3.10), as follows:

$$i_{dr,k}^{*} = -\frac{L_s}{1.5 L_m u_{ds,k}^{sv}} P_{s,k}^{*} \tag{4.1}$$

$$i_{qr,k}^{*} = \frac{L_s}{1.5 L_m u_{ds,k}^{sv}} Q_{s,k}^{*} - \frac{u_{ds,k}^{sv}}{\omega_{\bar{u}s,k} L_m} \tag{4.2}$$

The active voltage terms $\left(u_{dr,k}' \text{ and } u_{qr,k}'\right)$ provided by the PI current controllers and the compensating terms $\left(\Delta u_{dr,k}^{sv} \text{ and } \Delta u_{qr,k}^{sv}\right)$ computed in terms of generator variables are added to produce the reference d-q rotor voltage components in equations (3.19) and (3.20) that define the basic operation of SVOC.

DOI: 10.1201/9781003440529-5

Figure 4.1 provides a schematic for SVOC, in which the reference components of the rotor current can be calculated using the reference values of the active and reactive powers; then, they are compared with their actual components. The errors are then supplied to the PI current regulators to produce the active terms of the rotor voltage $\left(u'_{dr,k} \text{ and } u'_{qr,k} \right)$; then, a PLL [84, 85] is used to evaluate the stator voltage angle $\left(\theta_{\bar{u}_s} \right)$ which is necessary for the coordinates' transformation and also utilized in evaluating the slip frequency, which is required for evaluating the compensation voltage components $\Delta u^{sv}_{dr,k}$ and $\Delta u^{sv}_{qr,k}$. Eventually, the rotor voltage components $u^{sv}_{dr,k}$ and $u^{sv}_{qr,k}$ can be directly calculated after summing the compensation terms and active terms as mentioned in equations (3.19) and (3.20).

4.1.2 MPCC APPROACH

The principle of operation of the MPCC approach was described in a detailed manner in the previous chapter, in Section 3.2.2. The scheme of the MPCC is presented in Figure 4.2, in which the reference values of the rotor current can be evaluated at instant $(k+1)T_s$ after reformulating equations (4.1) and (4.2) as follows:

$$i^*_{dr,k+1} = -\frac{L_s}{1.5L_m u^{sv}_{ds,k+1}} P^*_{s,k+1} \tag{4.3}$$

$$i^*_{qr,k+1} = \frac{L_s}{1.5L_m u^{sv}_{ds,k+1}} Q^*_{s,k+1} - \frac{u^{sv}_{ds,k+1}}{\omega_{\bar{u}_s,k+1} L_m} \tag{4.4}$$

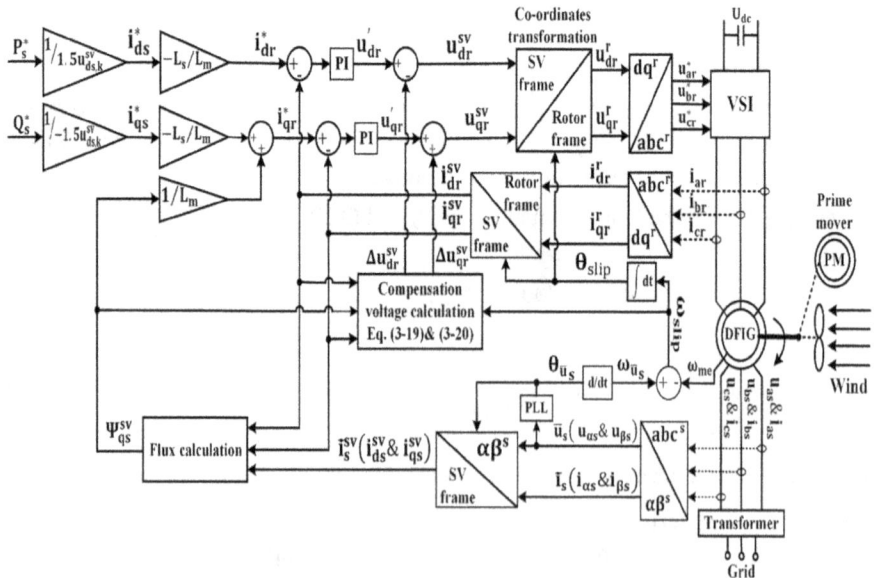

FIGURE 4.1   SVOC technique for the DFIG.

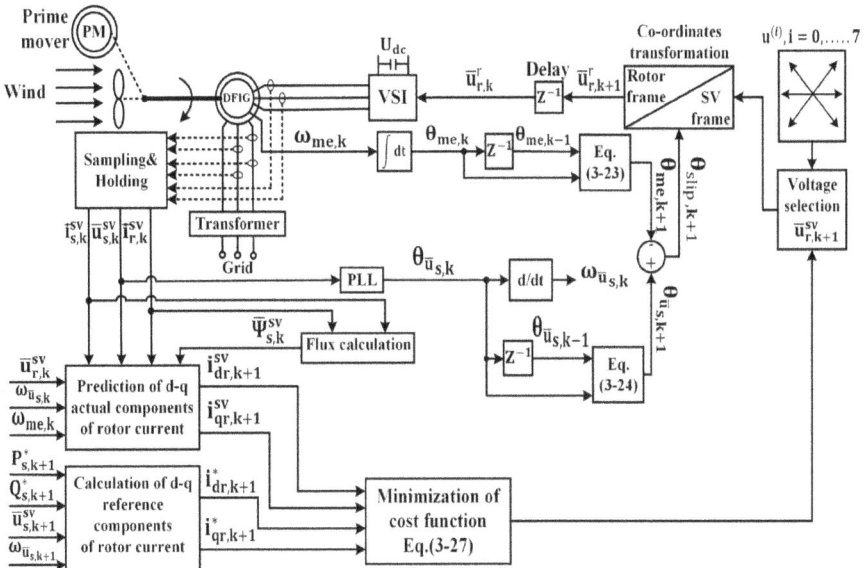

FIGURE 4.2 MPCC technique for the DFIG.

4.1.3 MPDTC APPROACH

The MPDTC technique was illustrated in detail in the previous chapter, in Section 3.2.3. Figure 4.3, introduces a schematic diagram for MPDTC technique in which the d-q components of the reference rotor current can be obtained using equations (4.3) and (4.4), meanwhile the reference stator current components can be evaluated in terms of reference rotor current components through equations (3.32) and (3.33), or via reference active and reactive powers as follows:

$$i^*_{ds,k+1} = \frac{P^*_{s,k+1}}{1.5u^{sv}_{ds,k+1}} \tag{4.5}$$

$$i^*_{qs,k+1} = \frac{Q^*_{s,k+1}}{-1.5u^{sv}_{ds,k+1}} \tag{4.6}$$

4.1.4 PROPOSED PVC APPROACH

The designed PVC uses a very simple cost function which consists of two similar components, which are the variances among the d-q reference components and d-q predicted components of the rotor voltage. The actual rotor voltage components $u^{sv}_{dr,k+1}$ and $u^{sv}_{qr,k+1}$ can be easily obtained using the switching states of the VSI, while the reference values of the rotor voltage $u^*_{dr,k+1}$ and $u^*_{qr,k+1}$ are obtained using two PI current regulators, which are designed as follows.

To design the PI rotor current regulators, the transfer function between the inputs, which are the rotor currents, and the outputs, which are the rotor voltages, must be

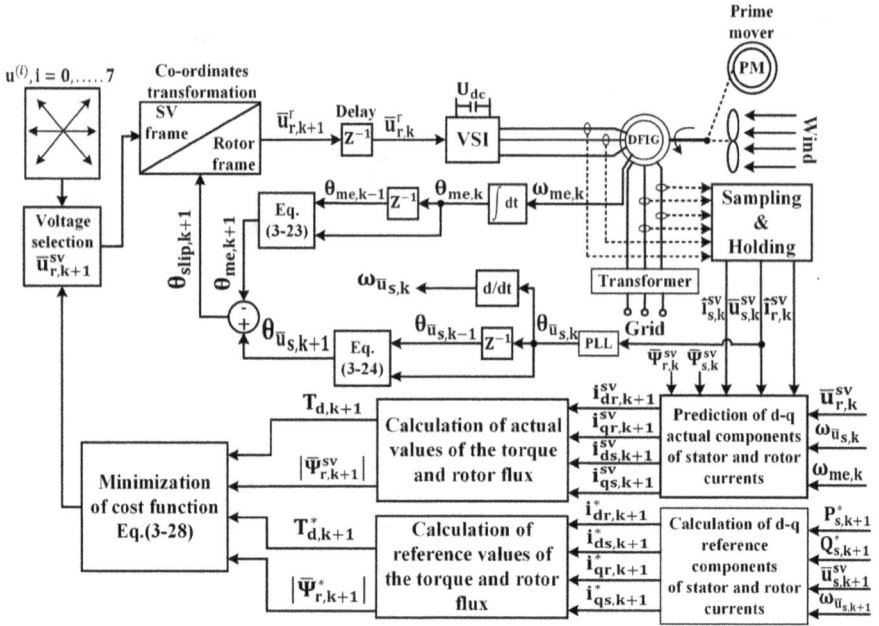

FIGURE 4.3 MPDTC technique for the DFIG.

formulated, and this can be performed under the SVOC principle and by reformulating equations (3.19) and (3.20) to be represented at instant $(k+1)T_s$ as follows:

$$u_{dr,k+1}^{sv} = \underbrace{R_r i_{dr,k+1}^{sv} + \sigma L_r \frac{di_{dr,k+1}^{sv}}{dt}}_{u'_{dr,k+1}\,=\,Active\ term} - \underbrace{\omega_{slip,k+1}\left(\frac{L_m}{L_s}\Psi_{qs,k+1}^{sv} + \sigma L_r i_{qr,k+1}^{sv}\right)}_{\Delta u_{dr,k+1}^{sv}\,=\,Compensation\ term} \qquad (4.7)$$

$$u_{qr,k+1}^{sv} = R_r i_{qr,k+1}^{sv} + \frac{L_m}{L_s}\frac{d\Psi_{qs,k+1}^{sv}}{dt} + \sigma L_r \frac{di_{qr,k+1}^{sv}}{dt} + \omega_{slip,k+1}\sigma L_r i_{dr,k+1}^{sv}$$

$$= \underbrace{R_r i_{qr,k+1}^{sv} + \sigma L_r \frac{di_{qr,k+1}^{sv}}{dt}}_{u'_{qr,k+1}\,=\,Active\ term} + \underbrace{\omega_{slip,k+1}\sigma L_r i_{dr,k+1}^{sv}}_{\Delta u_{qr,k+1}^{sv}\,=\,Compensation\ term} \qquad (4.8)$$

The actual components of the rotor current $i_{dr,k+1}^{sv}$ and $i_{qr,k+1}^{sv}$ are predicted using equations (3.25) and (3.26). In the same manner, $i_{qs,k+1}^{sv}$ can be predicted, which will then be utilized to evaluate $\Psi_{qs,k+1}^{sv}$ using equation (3.6) after being reformulated at instant $(k+1)T_s$.

The disturbances can be suppressed with the help of the compensation terms, while the active terms are used to achieve the desired change in the rotor current and drive the transfer function, which determines the response of the rotor voltage to any change in the rotor current. Then, the Laplace transform is applied to the active terms, assuming the initial current to be zero, resulting in the following:

$$u'_{dr,k+1}(s) = R_r i^{sv}_{dr,k+1}(s) + \sigma L_r \left[s i^{sv}_{dr,k+1}(s) - i^{sv}_{dr,k+1}(0) \right]$$

$$u'_{dr,k+1}(s) = R_r i^{sv}_{dr,k+1}(s) + \sigma L_r s i^{sv}_{dr,k+1}(s)$$

$$u'_{dr,k+1}(s) = i^{sv}_{dr,k+1}(s) \left[R_r + \sigma L_r s \right]$$

$$\frac{i^{sv}_{dr,k+1}(s)}{u'_{dr,k+1}(s)} = \frac{1}{R_r + \sigma L_r s} \tag{4.9}$$

In the same manner, the following formulation can be obtained:

$$\frac{i^{sv}_{qr,k+1}(s)}{u'_{qr,k+1}(s)} = \frac{1}{R_r + \sigma L_r s} \tag{4.10}$$

The transfer function of the PI rotor current regulators can also be represented by the following:

$$u'_{dr,k+1}(s) = \overbrace{\left(k_p + \frac{k_i}{s} \right)}^{PI} \overbrace{\left[i^*_{dr,k+1}(s) - i^{sv}_{dr,k+1}(s) \right]}^{d - current\ error} \tag{4.11}$$

$$u'_{qr,k+1}(s) = \underbrace{\left(k_p + \frac{k_i}{s} \right)}_{PI} \underbrace{\left[i^*_{qr,k+1}(s) - i^{sv}_{qr,k+1}(s) \right]}_{q - current\ error} \tag{4.12}$$

In equations (4.9)–(4.12), the transfer function of both PI rotor current regulators is similar, which can be derived as follows.

By dividing both sides of equations (4.11) and (4.12) on $i^{sv}_{dr,k+1}(s)$ and $i^{sv}_{qr,k+1}(s)$, respectively, we obtain:

$$\frac{u'_{dr,k+1}(s)}{i^{sv}_{dr,k+1}(s)} = \left(\frac{k_p s + k_i}{s} \right) \left[\frac{i^*_{dr,k+1}(s)}{i^{sv}_{dr,k+1}(s)} - 1 \right] \tag{4.13}$$

$$\frac{u'_{qr,k+1}(s)}{i^{sv}_{qr,k+1}(s)} = \left(\frac{k_p s + k_i}{s} \right) \left[\frac{i^*_{qr,k+1}(s)}{i^{sv}_{qr,k+1}(s)} - 1 \right] \tag{4.14}$$

Then, by substituting the voltage component $\{u'_{dr,k+1}(s)\}$ from equation (4.9) into equation (4.13), it results in the following:

$$\frac{i^{sv}_{dr,k+1}(s)}{i^{sv}_{dr,k+1}(s)} (R_r + \sigma L_r s) = \left(\frac{k_p s + k_i}{s} \right) \left[\frac{i^*_{dr,k+1}(s)}{i^{sv}_{dr,k+1}(s)} - 1 \right]$$

$$R_r + \sigma L_r s = \frac{k_p s + k_i}{s} \left| \frac{i^*_{dr,k+1}(s)}{i^{sv}_{dr,k+1}(s)} \right| - \frac{k_p s + k_i}{s}$$

$$\frac{k_p s + k_i}{s} \left| \frac{i^*_{dr,k+1}(s)}{i^{sv}_{dr,k+1}(s)} \right| = R_r + \sigma L_r s + \frac{k_p s + k_i}{s}$$

$$\frac{i^*_{dr,k+1}(s)}{i^{sv}_{dr,k+1}(s)} = \left(\frac{R_r s + \sigma L_r s^2 + k_p s + k_i}{s} \right) * \frac{s}{k_p s + k_i}$$

$$\frac{i^{sv}_{dr,k+1}(s)}{i^*_{dr,k+1}(s)} = \frac{k_p s + k_i}{\sigma L_r s^2 + (R_r + k_p)s + k_i} \qquad (4.15)$$

In the same manner, we can obtain:

$$\frac{i^{sv}_{qr,k+1}(s)}{i^*_{qr,k+1}(s)} = \frac{k_p s + k_i}{\sigma L_r s^2 + (R_r + k_p)s + k_i} \qquad (4.16)$$

The denominator of both equations (4.15) and (4.16) is considered as the characteristic equation which controls the dynamics of the PI current regulators, and this equation should have negative real roots to achieve system stability; then, the following must be accomplished:

$$\sigma L_r s^2 + (R_r + k_p)s + k_i = 0 \qquad (4.17)$$

By dividing both sides of equation (4.17) by (σL_r), we obtain:

$$s^2 + \frac{(R_r + k_p)}{\sigma L_r} s + \frac{k_i}{\sigma L_r} = 0 \qquad (4.18)$$

Then, to calculate the parameters k_p and k_i of the PI rotor current regulators, we must compare the terms of equation (4.18) with that of the characteristic equation of the second-order system, which is represented by:

$$s^2 + 2D\omega_n s + \omega_n^2 = 0 \qquad (4.19)$$

Finally, by comparing equations (4.18) and (4.19), the parameters of the PI current regulators can be calculated as follows:

$$k_p = 2D\omega_n \sigma L_r - R_r \quad \text{and} \quad k_i = \sigma L_r \omega_n^2 \qquad (4.20)$$

After completing the design of the PI current regulators, the reference components of the rotor voltage $u^*_{dr,k+1}$ and $u^*_{qr,k+1}$ can be obtained and then utilized by the cost function.

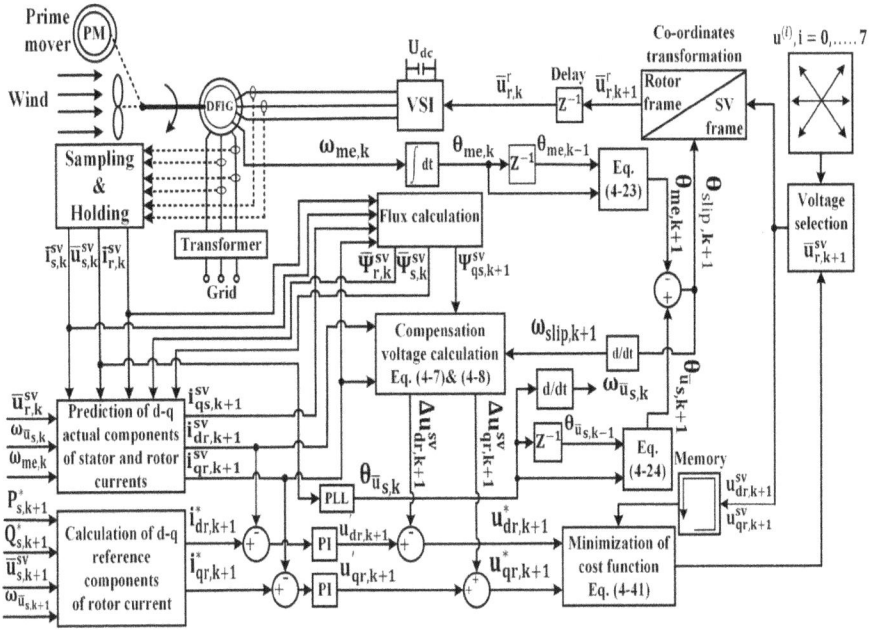

FIGURE 4.4 Proposed PVC technique for the DFIG.

Figure 4.4 presents the scheme of the designed PVC, which starts with measuring the stator voltage and stator current and then sampling these signals. As mentioned previously, Taylor expansion is used to predict the actual components of the stator and rotor current, as stated in equations (3.25) and (3.26). The actual rotor current components $i_{dr,k+1}^{sv}$ and $i_{qr,k+1}^{sv}$ are then compared with the reference d-q rotor current components $i_{dr,k+1}^*$ and $i_{qr,k+1}^*$, which can be evaluated using equations (4.3) and (4.4), respectively. After that, the errors of the rotor current components are fed to the two designed PI current regulators to obtain the active term components of the rotor voltage $u'_{dr,k+1}$ and $u'_{qr,k+1}$, which are then added to the compensation voltage components $\Delta u_{dr,k+1}^{sv}$ and $\Delta u_{qr,k+1}^{sv}$ to obtain the reference d-q components of the rotor voltage $u_{dr,k+1}^*$ and $u_{qr,k+1}^*$ as stated in equations (4.7) and (4.8), respectively. As mentioned earlier, the actual components of the rotor voltage $u_{dr,k+1}^{sv}$ and $u_{qr,k+1}^{sv}$ are directly obtained using the switching states of the inverter. Finally, the actual and reference rotor voltage components are utilized by the cost function (3.41). As noted in equation (3.41), there is no need for a weighting factor, which leads to avoiding the wrong selection, which may occur with MPDTC. Furthermore, it does not involve any estimated variables that depend on the machine parameters as in MPCC and MPDTC. All of these modifications make PVC more robust and more suitable to be utilized.

4.2 COMPUTATION RESULTS

Tests were carried out by using MATLAB simulation (Simulink).

4.2.1   EVALUATION THE PERFORMANCE OF THE GRID-CONNECTED DFIG UNDER WIND-DRIVEN OPERATION

The performance of the DFIG was studied with SVOC, MPCC, MPDTC, and with the designed predictive controller. The DFIG was tested under different operating speeds to prove its ability to provide the required load power and keep the output voltage and frequency at constant values regardless of the wind variations. The wind speed variation is presented in Figure 4.5, meanwhile the driven speeds of the DFIG are introduced in Figure 4.6. The data of the DFIG and the wind turbine are presented in Appendix A, in Table A1, meanwhile the control system parameters are introduced in Table A5.

4.2.1.1   Testing with SVOC Strategy

The DFIG's performance was studied with the SVOC algorithm under the stated operating speeds and the obtained results reveal that the active power, reactive power, developed torque, and rotor flux follow their reference values smoothly as presented in Figures 4.7–4.9. In addition, the stator and rotor currents track the power change as noticed in Figures 4.10 and 4.11. It can be deduced that the SVOC approach is free of ripples, and the actual values of the powers, torque, and rotor flux are tracking their references. The main defects are the complexity of the control system and the delay of its dynamic response which is caused by the PI controllers.

4.2.1.2   Testing with MPCC Strategy

The performance of the DFIG was studied with the MPCC approach for the same operating conditions which was stated in Section 3.3.1. The results are shown in

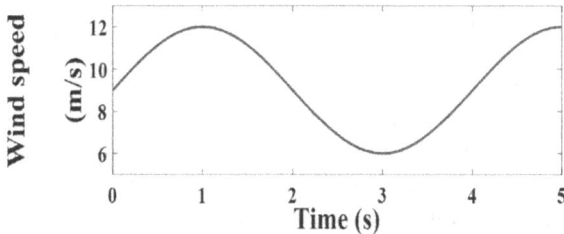

FIGURE 4.5   Wind speed variation (m/s).

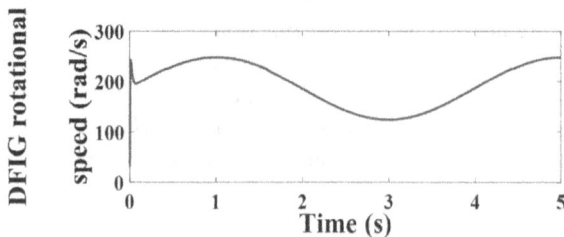

FIGURE 4.6   Operating speeds of the DFIG (rad/s).

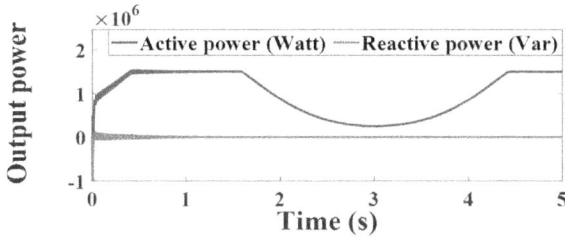

FIGURE 4.7   Output power under SVOC.

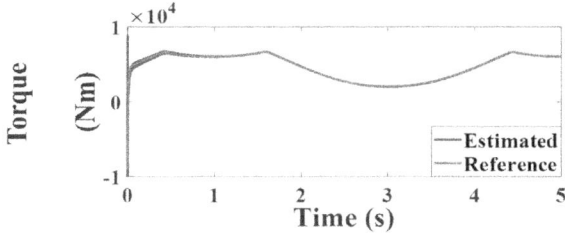

FIGURE 4.8   Torque developed under SVOC under wind driven operation (Nm).

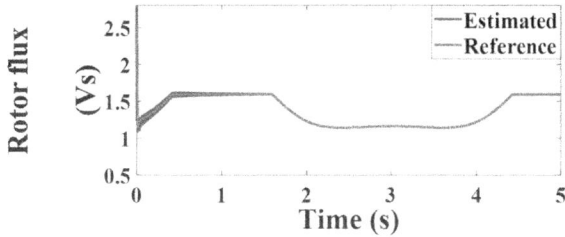

FIGURE 4.9   Rotor flux under SVOC under wind driven operation (Vs).

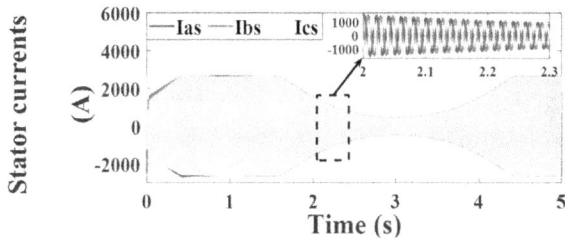

FIGURE 4.10   Stator currents under SVOC under wind driven operation (A).

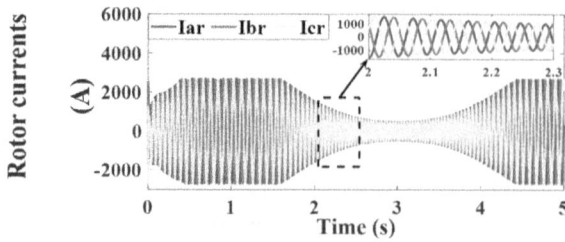

FIGURE 4.11   Rotor currents under SVOC under wind driven operation (A).

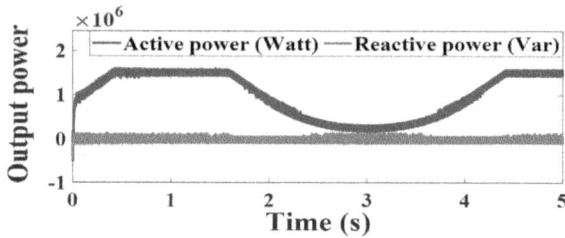

FIGURE 4.12   Output power under MPCC.

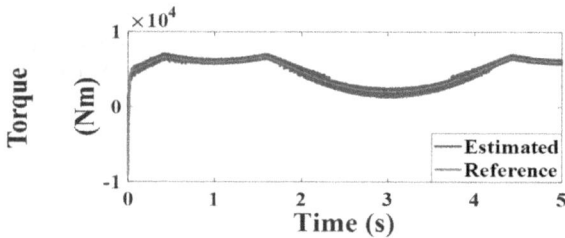

FIGURE 4.13   Torque developed under MPCC (Nm) under wind driven operation.

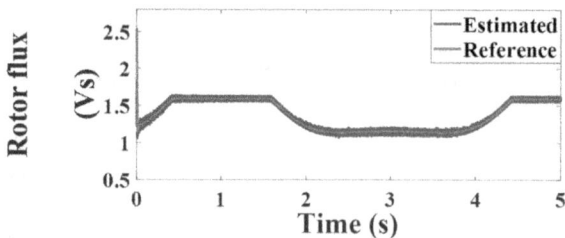

FIGURE 4.14   Rotor flux under MPCC (Vs) under wind driven operation.

Figures 4.12–4.16, which clarify that the actual values follow their references with a faster dynamic response than that of the SVOC; unfortunately, its ripples' content is remarkable compared with that of the SVOC scheme.

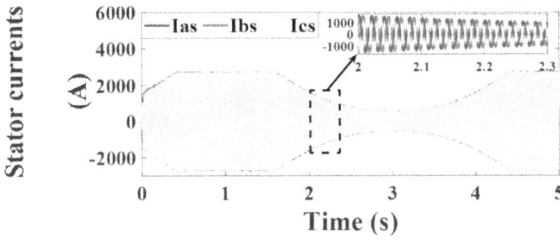

FIGURE 4.15 Stator currents under MPCC (A) under wind driven operation.

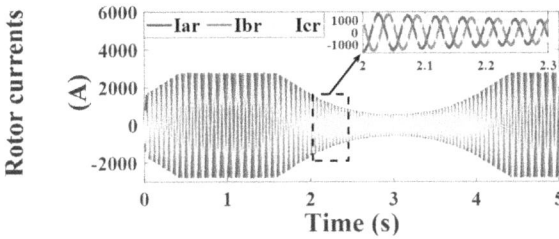

FIGURE 4.16 Rotor currents under MPCC (A) under wind driven operation.

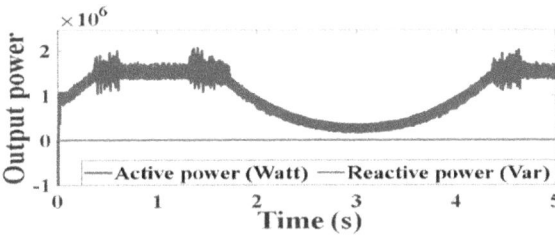

FIGURE 4.17 Output power under MPDTC.

4.2.1.3 Testing with MPDTC Strategy

The performance of the DFIG was tested under MPDTC approach to be able to compare it with that of the previous controllers and the proposed predictive controller. The results related to MPDTC are shown in Figures 4.17–4.21.

4.2.1.4 Testing with Proposed PVC Strategy

The DFIG performance is tested under the proposed predictive control strategy through the various wind speeds and the results are presented in Figures 4.22–4.26. The results reveal that the actual values of active, reactive, torque and flux track effectively their references. Furthermore, the ripples content is greatly suppressed compared with MPDTC and MPCC techniques.

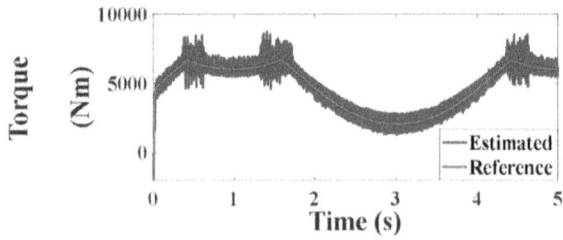

FIGURE 4.18   Torque developed under MPDTC (Nm) under wind driven operation.

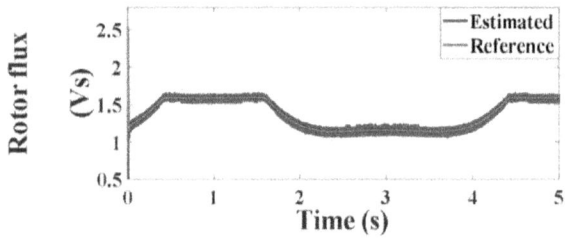

FIGURE 4.19   Rotor flux under MPDTC (Vs) under wind driven operation.

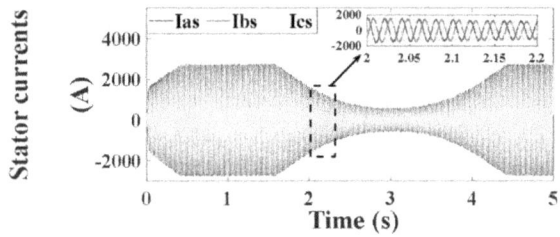

FIGURE 4.20   Stator currents under MPDTC under wind driven operation (A).

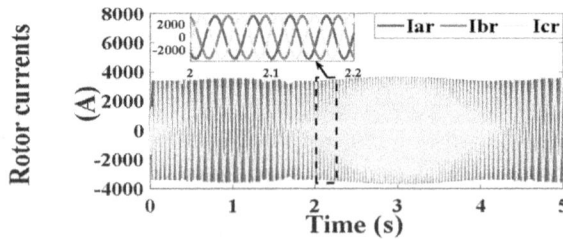

FIGURE 4.21   Rotor currents under MPDTC (A) under wind driven operation.

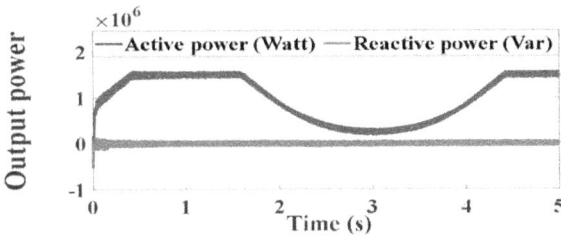

FIGURE 4.22 Output power under proposed predictive controller.

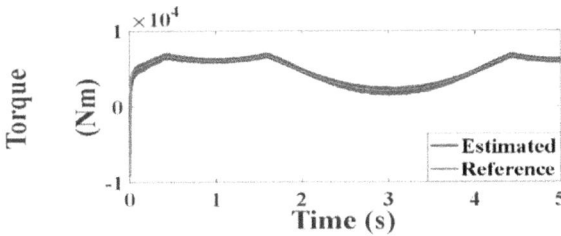

FIGURE 4.23 Developed torque under proposed predictive controller (Nm).

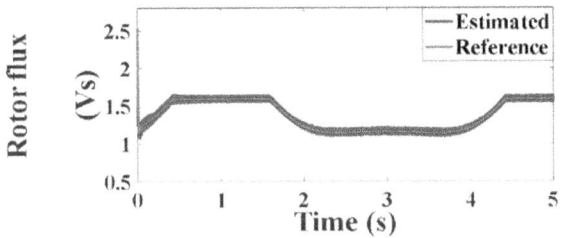

FIGURE 4.24 Rotor flux under proposed predictive controller (Vs).

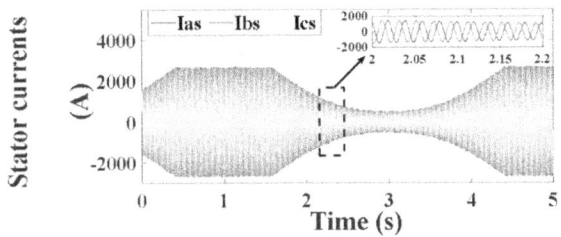

FIGURE 4.25 Stator currents under proposed predictive controller (A).

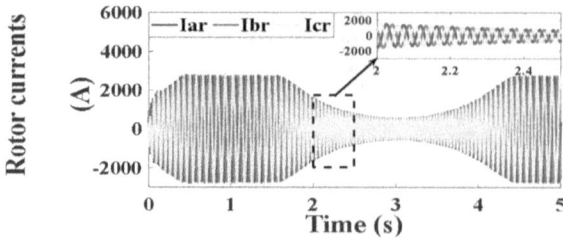

FIGURE 4.26 Rotor currents under proposed predictive controller (A).

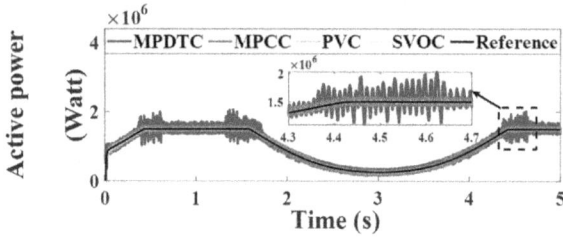

FIGURE 4.27 Active power (Watt).

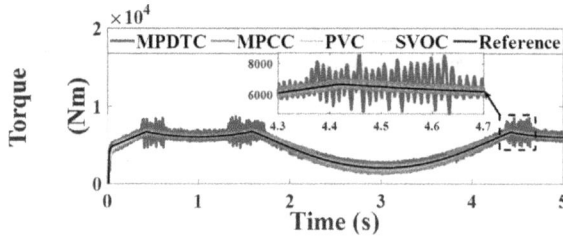

FIGURE 4.28 Developed torque (Nm).

4.2.1.5 Comparison Study

Eventually, a detailed comparison between the dynamic performance of the DFIG under SVOC, MPCC, MPDTC and the designed predictive control approaches is performed, while showing the merits and defects of each strategy, to identify the most convenient method to be adopted with the DFIG. The comparison is performed in terms of ripples content, number of performed commutations and THD. Figures 4.27–4.29 introduce the results of comparing the dynamic performance of the four approaches, which clarify that the formulated predictive control scheme has a response faster than that of the SVOC, MPCC, and MPDTC algorithms. Moreover, the proposed approach introduces less ripples when compared with the MPDTC and MPCC; this is also confirmed by the numerical values of Table 4.1 and Figure 4.30. Table 4.2 introduces a unique comparison among the predictive controllers, as it compares the number of commutations performed by the three predictive controllers to easily determine which strategy has lower computational burden. The designed

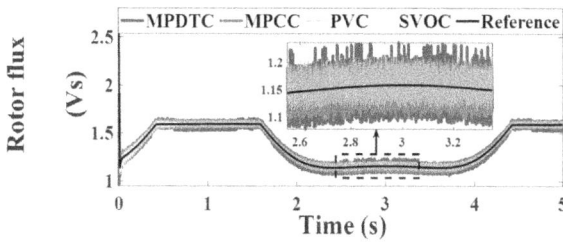

FIGURE 4.29   Rotor flux (Vs).

TABLE 4.1
Ripples' Content of the Actual Values above Their References

Algorithm	Ripples of Active Power (Watt)	Ripples of Developed Torque (Nm)	Ripples of Rotor Flux (Vs)
SVOC	22,000	109	0.018
MPCC	84,000	935	0.095
MPDTC	93,100	1399	0.108
PVC	53,900	397	0.084

predictive scheme introduces lower computational burden, as its number of commutations are lower compared with the MPDTC and MPCC, as also presented in Figure 4.31.

Figures 4.32–4.40, introduce the spectrums of the stator current phases under the MPCC, MPDTC, and proposed predictive scheme, the THD of the proposed control is lower than that of MPCC and MPDTC as shown from the comparison of the spectrums. This fact is also confirmed by the numerical values of Table 4.3 and the chart of Figure 4.41.

Lastly, it can be deduced that the proposed predictive control is the most appropriate technique to be used with the DFIG, as it can overcome most shortages of the other controller. Compared with MPDTC, the formulated controller has faster dynamic response, lower ripples, less computation burden and lower THD.

4.2.2   EVALUATION THE PERFORMANCE OF THE GRID-CONNECTED DFIG UNDER DIRECT-DRIVEN OPERATION

Tests were carried out for the four control techniques (SVOC, MPCC, MPDTC, and proposed PVC) under two different operating conditions. In the first one, the active and reactive power reference values $\left(P_{s,k}^{*}\ \text{and}\ Q_{s,k}^{*}\right)$ were kept constant and set to 50 kW and 0.0 var, respectively, while the DFIG was driven by various wind-driven operating speeds (sub-synchronous $= -30\%$ of $\omega_{\bar{u}_{s,k}}$; synchronous $= \omega_{\bar{u}_{s,k}}$; super-synchronous $=+30\%$ of $\omega_{\bar{u}_{s,k}}$), as shown in Figure 4.42. The DFIG was driven by different operating speeds to ensure the control's ability to maintain the desired power

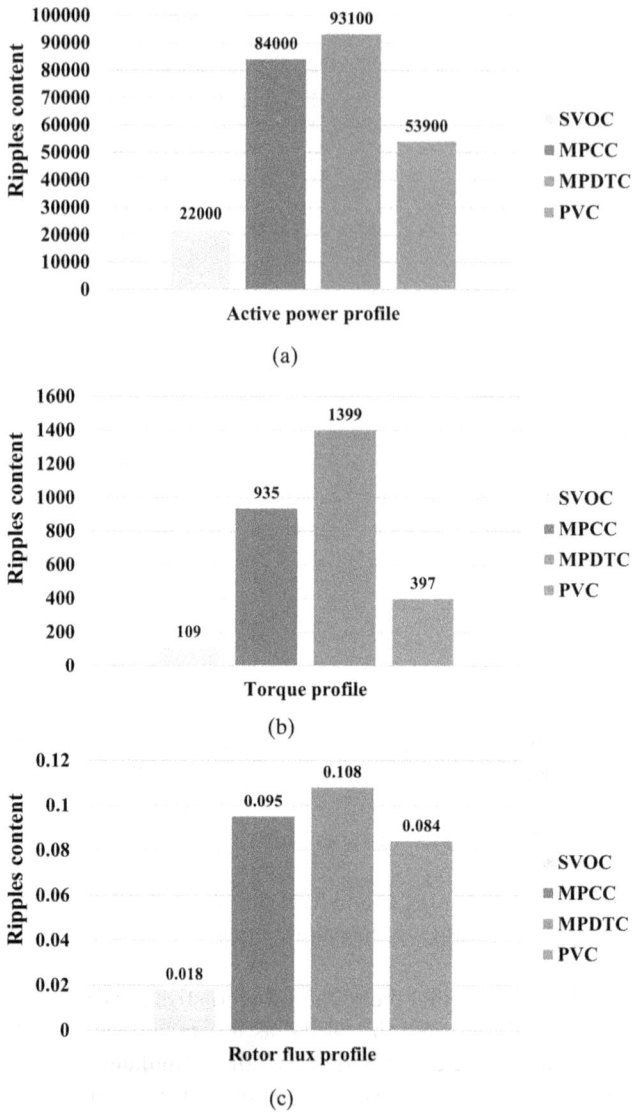

FIGURE 4.30 Histogram of ripples' content of the actual values above their references. (a) Active power profile (Watt). (b) Torque profile (Nm). (c) Rotor flux profile (Vs).

TABLE 4.2

Comparison in Terms of Performed
Commutations by the Predictive
Controllers with variable operating speed.

Technique	No. of Commutations
MPCC	7,008
MPDTC	6,772
Proposed PVC	3,465

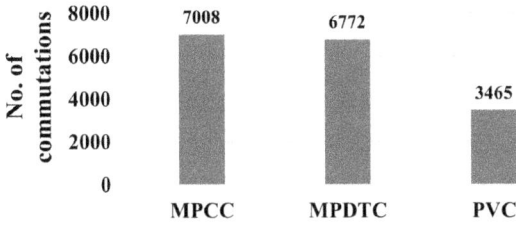

FIGURE 4.31 Histogram of number of performed commutations by the predictive controllers under wind driven operation.

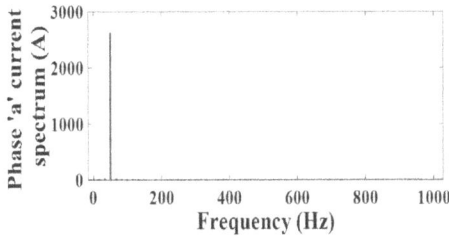

FIGURE 4.32 THD of Phase "a" of stator current under MPCC.

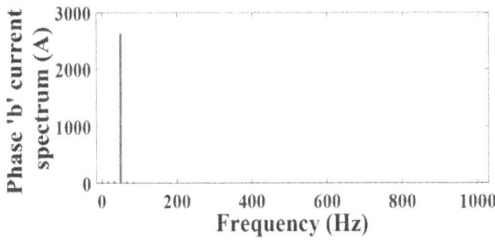

FIGURE 4.33 THD of Phase "b" of stator current under MPCC.

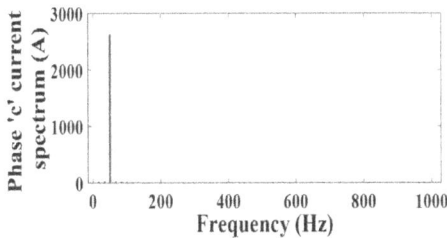

FIGURE 4.34 THD of Phase "c" of stator current under MPCC.

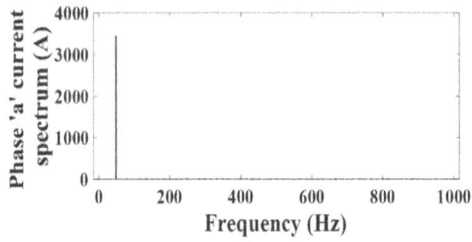

FIGURE 4.35   THD of Phase "a" of stator current under MPDTC.

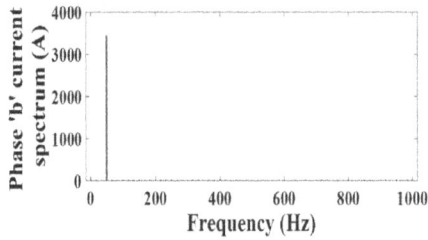

FIGURE 4.36   THD of Phase "b" of stator current under MPDTC.

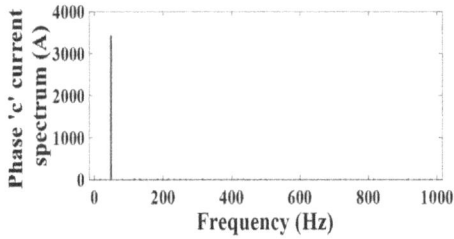

FIGURE 4.37   THD of Phase "c" of stator current under MPDTC.

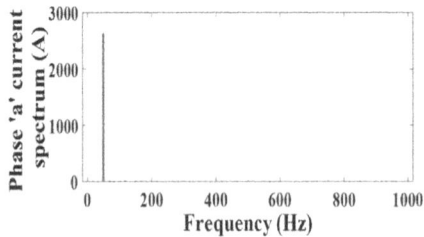

FIGURE 4.38   THD of Phase "a" of stator current under proposed controller.

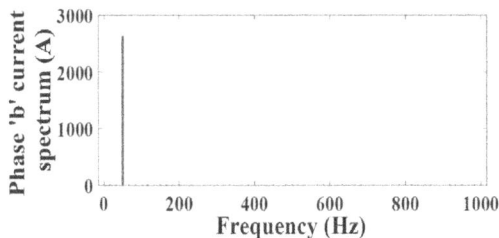

FIGURE 4.39 THD of Phase "b" of stator current under proposed controller.

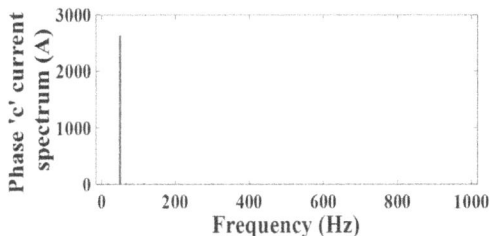

FIGURE 4.40 THD of Phase "c" of stator current under proposed controller.

TABLE 4.3
FFT Analysis for Components of the Stator Current

		Phase A	Phase B	Phase C
MPCC	Fundamental	2625.53 A	2628.1 A	2625.92 A
	THD	0.70%	0.49%	0.65%
MPDTC	Fundamental	3442.89 A	3440.24 A	3431.84 A
	THD	0.76%	0.88%	0.73%
PVC	Fundamental	2626.83 A	2628.59 A	2635.07 A
	THD	0.59%	0.46%	0.58%

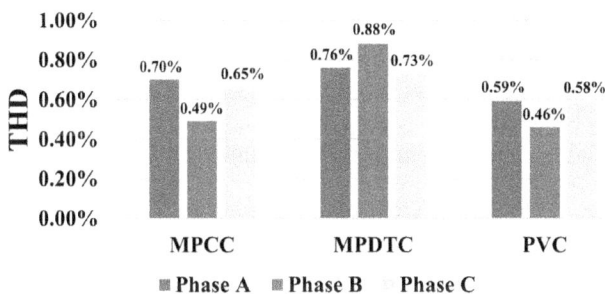

FIGURE 4.41 Histogram of THD analysis for the stator current components under wind driven operation.

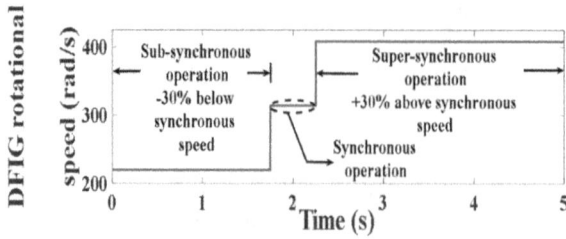

FIGURE 4.42   Prime mover operating speeds (rad/s) under direct driven operation.

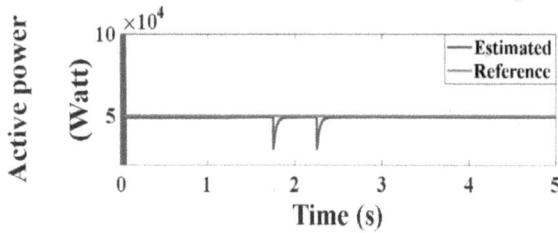

FIGURE 4.43   Active power under SVOC (Watt) under direct driven operation.

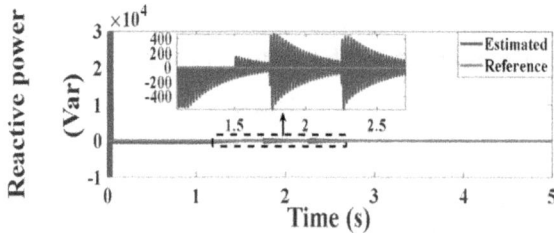

FIGURE 4.44   Reactive power under SVOC (Var) under direct driven operation.

ratings regardless of the speed variations. Meanwhile, in the second condition, the reference power was initially set to 25 kW and then changed to 50 kW while maintaining the reactive power at 0.0 var. Regarding the wind-driven speed, it was kept constant and set to a synchronous speed $\left(\omega_{\bar{u}_s,k} = 314 \text{ rad/s}\right)$, as shown in Figure 4.49. The parameters of DFIG are presented in Appendix A, in Table A1, while the data specification of the control system is listed in Table A5.

4.2.2.1   Testing with SVOC Technique

The DFIG's performance was tested with the SVOC technique for two different operating regimes: the first under speed variation as shown in Figure 4.42 and the other under a fixed speed of 314 rad/s as presented in Figure 3.49. The results obtained for both conditions, which are shown in Figures 4.43–4.48 for the variable speed and constant active power (Figure 4.49) and in Figures 4.50–4.53 for the fixed speed and variable active power profile, demonstrate that the active power, reactive power,

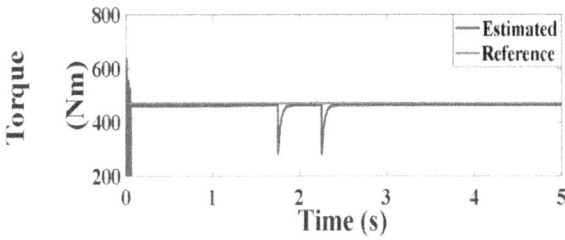

FIGURE 4.45   Torque developed under SVOC (Nm).

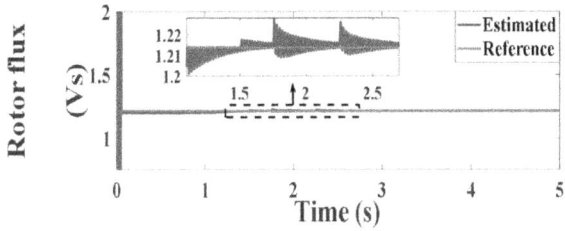

FIGURE 4.46   Rotor flux under SVOC (Vs).

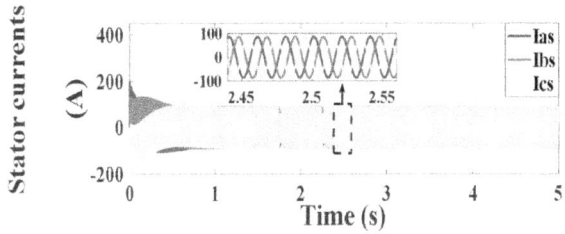

FIGURE 4.47   Stator currents under SVOC (A).

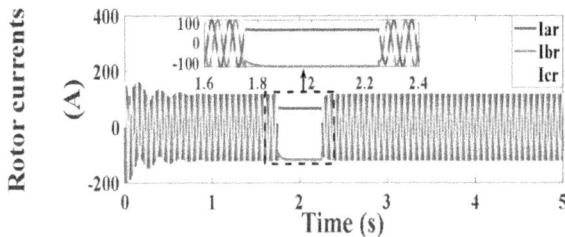

FIGURE 4.48   Rotor currents under SVOC (A).

developed torque, and rotor flux smoothly follow their reference values. In addition, the stator and rotor currents track the power change as shown in Figures 4.54 and 4.55. It can be concluded that the SVOC is ripple-free, and the estimated values of

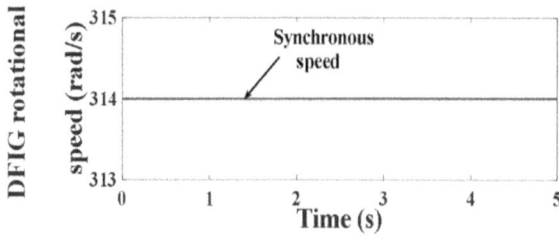

FIGURE 4.49   Prime mover operating speed (rad/s) under direct driven operation.

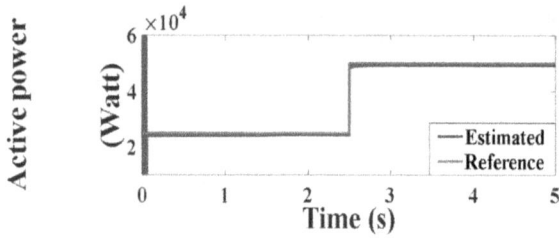

FIGURE 4.50   Active power under SVOC (Watt).

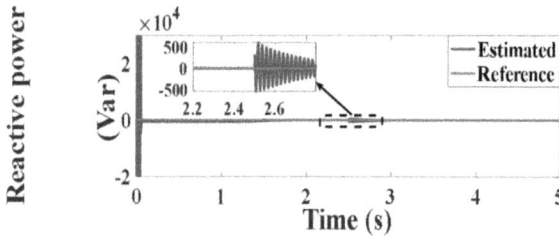

FIGURE 4.51   Reactive power under SVOC (Var).

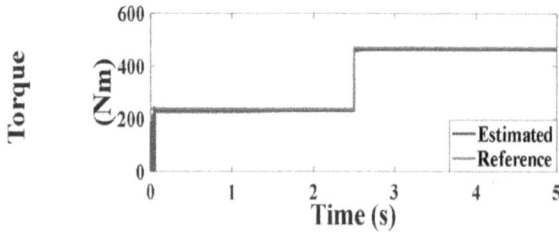

FIGURE 4.52   Torque developed under SVOC (Nm).

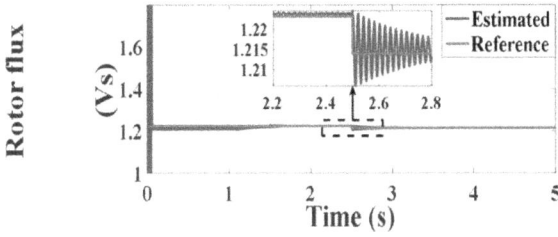

FIGURE 4.53   Rotor flux under SVOC (Vs).

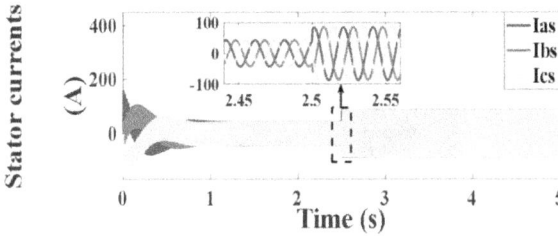

FIGURE 4.54   Stator currents under SVOC (A).

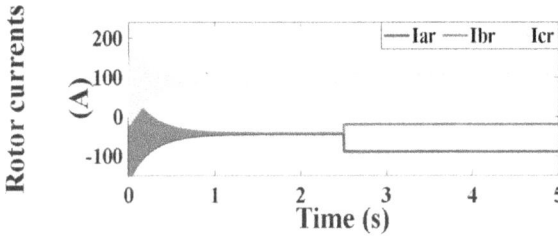

FIGURE 4.55   Rotor currents under SVOC (A).

the powers, torque, and rotor flux are tracking their references in a good manner. The main shortages are the control system complexity and the delayed dynamic response caused by the current regulators.

4.2.2.2 Testing with MPCC Technique

The DFIG's performance was tested with the MPCC technique for the same operating conditions presented in Section 4.3.2.1. The results are shown in Figures 4.56–4.62 for variable speed operation and in Figures 4.63–4.69 for fixed speed operation. The captured results indicate that the actual values of the active and reactive powers, torque, and rotor flux follow their references with a dynamic response faster than that of the SVOC principle; however, its ripples are remarkable in comparison with the SVOC technique.

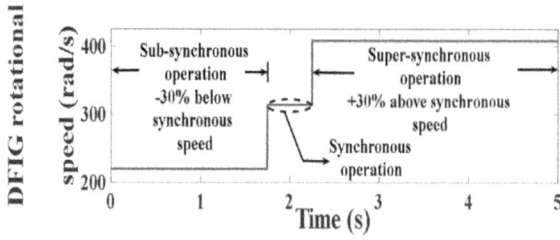

FIGURE 4.56   Prime mover operating speeds (rad/s).

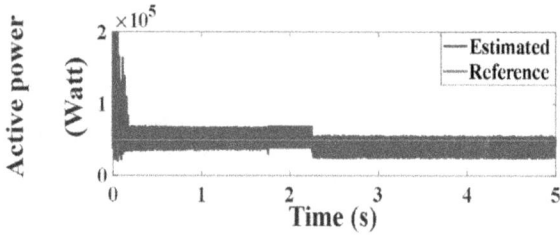

FIGURE 4.57   Active power under MPCC under direct driven operation (Watt).

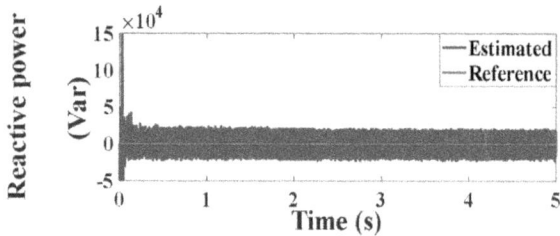

FIGURE 4.58   Reactive power under MPCC (Var) under direct driven operation.

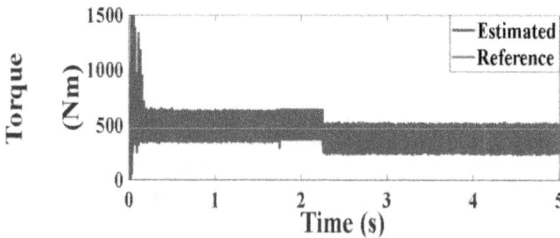

FIGURE 4.59   Torque developed under MPCC (Nm).

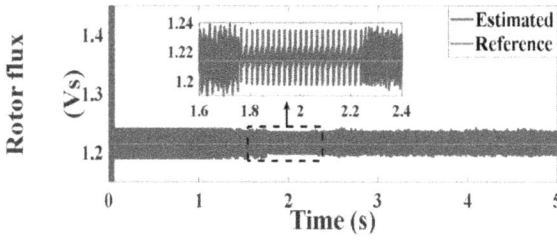

FIGURE 4.60   Rotor flux under MPCC (Vs).

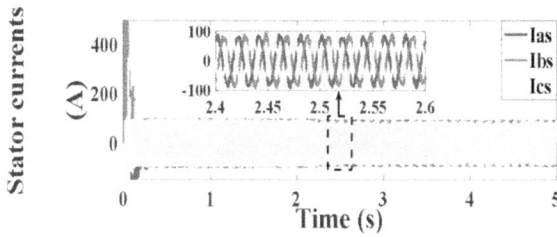

FIGURE 4.61   Stator currents under MPCC (A).

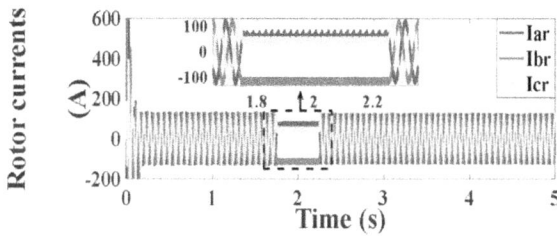

FIGURE 4.62   Rotor currents under MPCC (A).

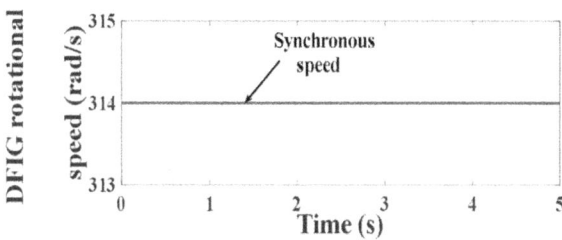

FIGURE 4.63   Prime mover operating speed (rad/s).

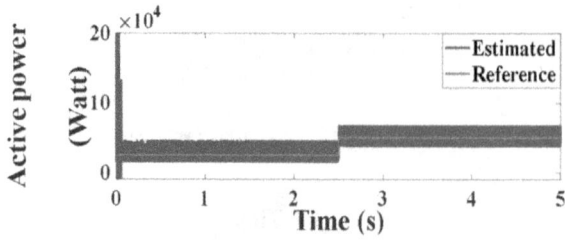

FIGURE 4.64   Active power under MPCC (Watt).

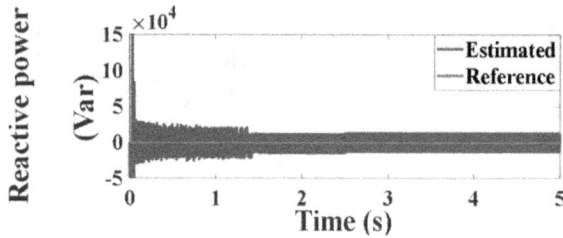

FIGURE 4.65   Reactive power under MPCC (Var).

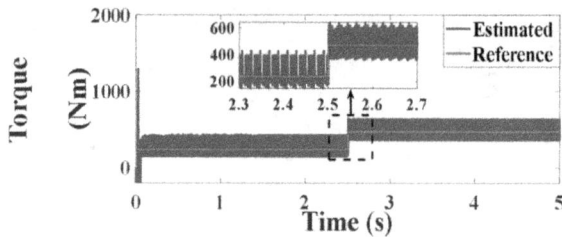

FIGURE 4.66   Torque developed under MPCC (Nm).

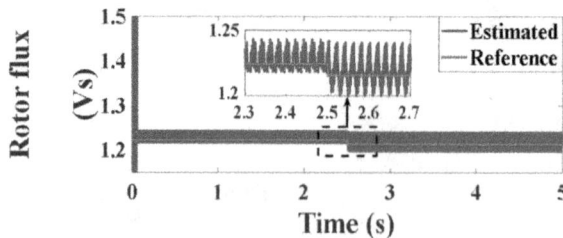

FIGURE 4.67   Rotor flux under MPCC (Vs).

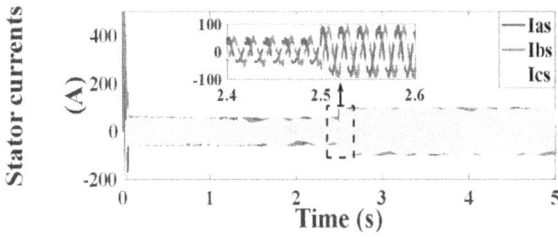

FIGURE 4.68 Stator currents under MPCC (A).

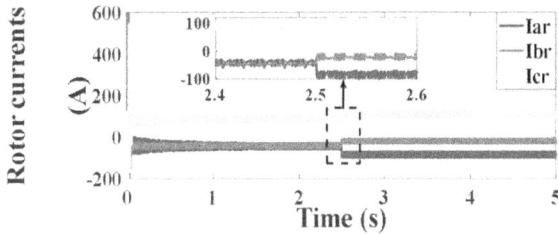

FIGURE 4.69 Rotor currents under MPCC (A).

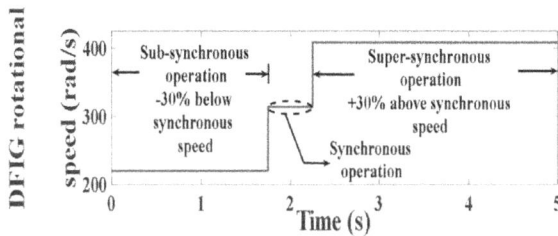

FIGURE 4.70 Prime mover operating speeds (rad/s).

4.2.2.3 Testing with MPDTC Technique

The DFIG's performance was tested with the MPDTC technique, and the results for both operating conditions are, respectively, shown in Figures 4.70–4.83, which illustrate that the actual values of the powers, torque, and rotor flux follow their references with a dynamic response faster than that of the MPCC and SVOC principles, but unfortunately, it has more ripples than the MPCC and SVOC techniques.

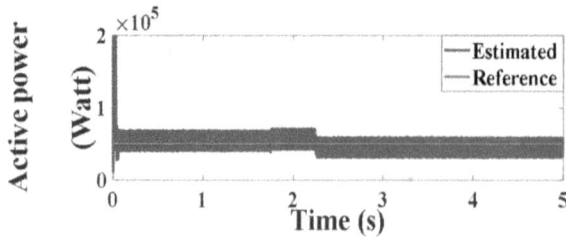

FIGURE 4.71   Active power under MPDTC under direct driven operation (Watt).

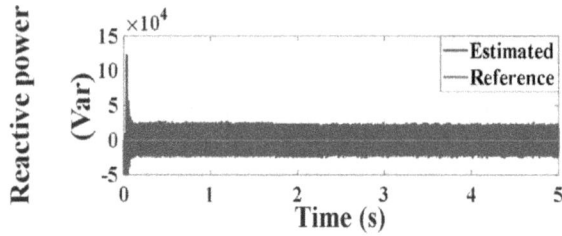

FIGURE 4.72   Reactive power under MPDTC under direct driven operation (Var).

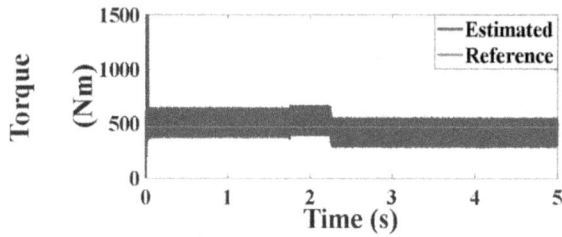

FIGURE 4.73   Torque under MPDTC (Nm) under direct driven operation.

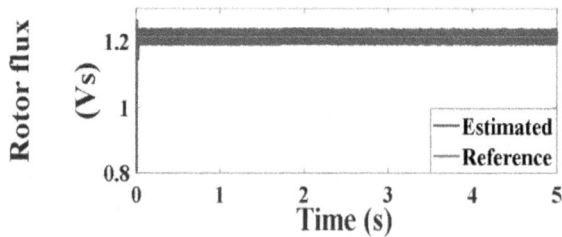

FIGURE 4.74   Rotor flux under MPDTC (Vs).

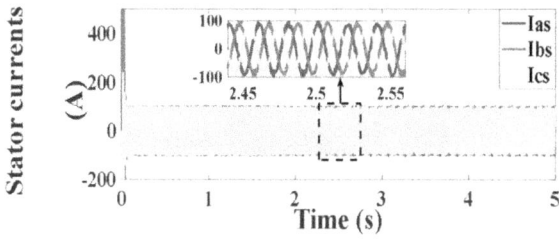

FIGURE 4.75   Stator currents under MPDTC (A).

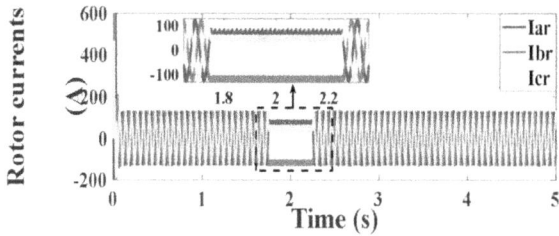

FIGURE 4.76   Rotor currents under MPDTC (A).

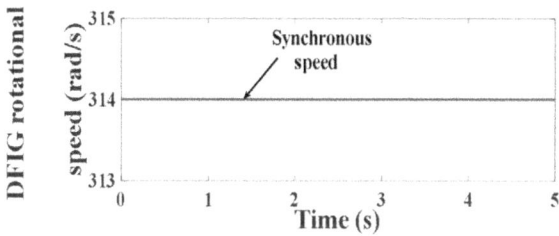

FIGURE 4.77   Prime mover operating speed (rad/s).

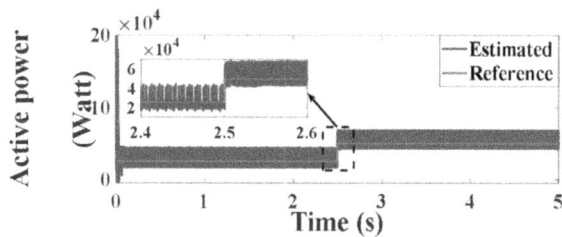

FIGURE 4.78   Active power under MPDTC (Watt).

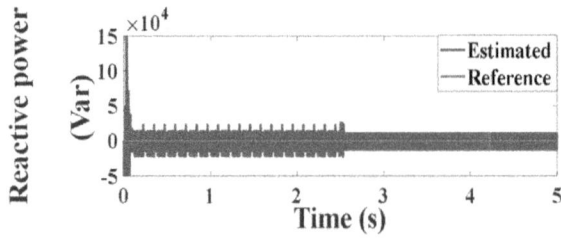

FIGURE 4.79   Reactive power under MPDTC (Var).

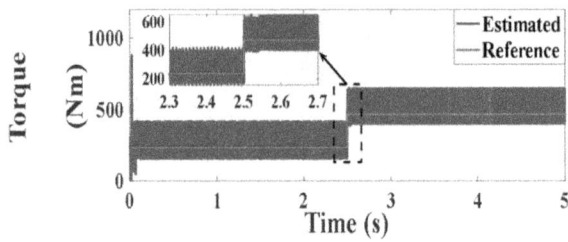

FIGURE 4.80   Torque under MPDTC (Nm).

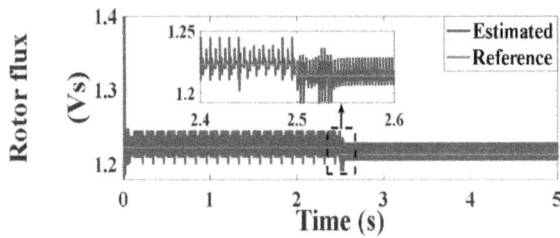

FIGURE 4.81   Rotor flux under MPDTC (Vs).

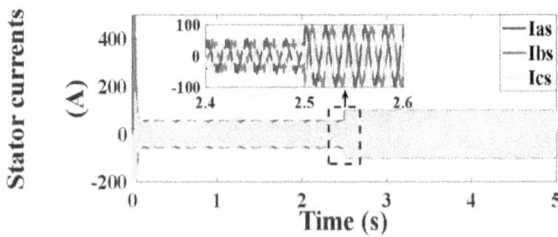

FIGURE 4.82   Stator currents under MPDTC (A).

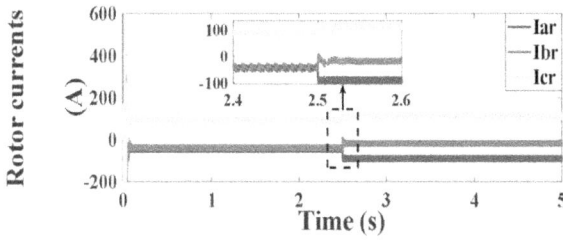

FIGURE 4.83 Rotor currents under MPDTC (A).

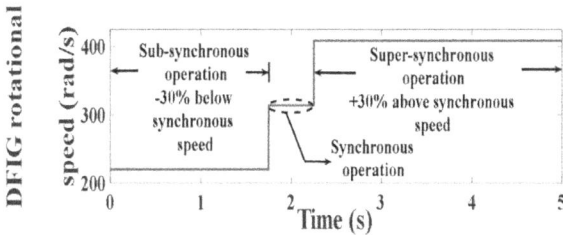

FIGURE 4.84 Prime mover operating speeds (rad/s).

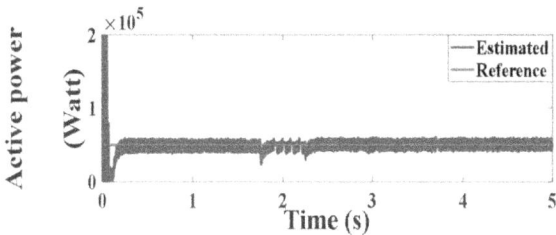

FIGURE 4.85 Active power under PVC (Watt) under direct driven operation.

4.2.2.4 Testing with Proposed PVC Technique

The DFIG's performance was tested with the proposed PVC technique, and the results are shown in Figures 4.84–4.90 for the variable speed operation and Figures 4.91–4.97 for the fixed speed operation. These results prove and confirm that the proposed PVC control system has successfully achieved its targets, as the actual estimated values of the powers, torque developed, and rotor flux follow their references with high precision in the two conditions. Furthermore, the ripples' content is effectively suppressed compared with the values under MPDTC and MPCC. In addition, the dynamic response of PVC is the fastest in comparison with that of the MPDTC, MPCC, and SVOC techniques.

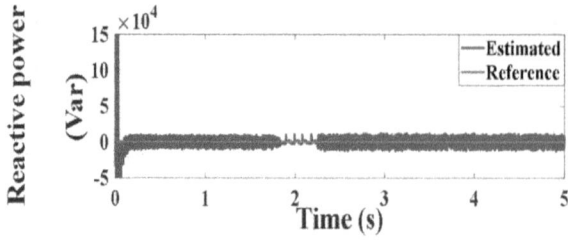

FIGURE 4.86   Reactive power under PVC (Var) under direct driven operation.

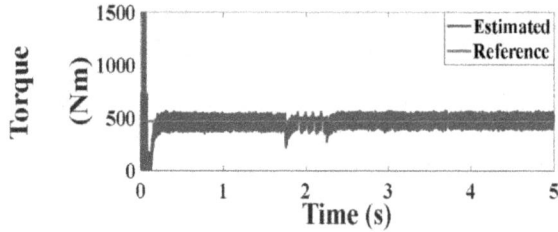

FIGURE 4.87   Torque developed with PVC (Nm) under direct driven operation.

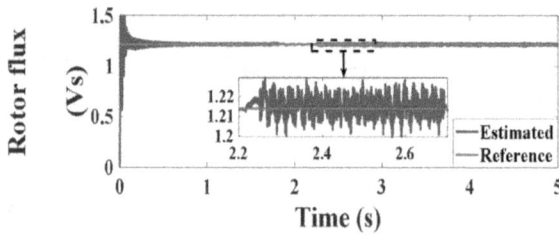

FIGURE 4.88   Rotor flux with PVC (Vs) under direct driven operation.

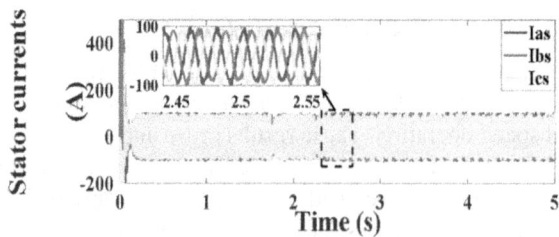

FIGURE 4.89   Stator currents with PVC (A).

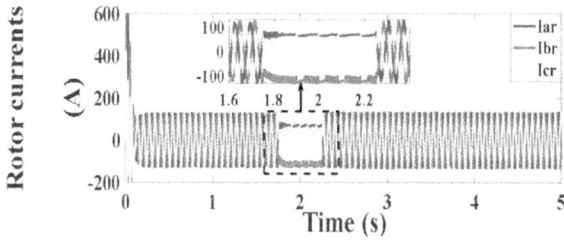

FIGURE 4.90 Rotor currents with PVC (A).

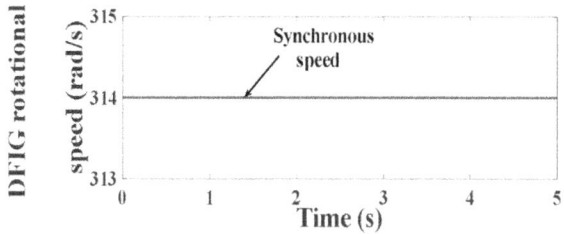

FIGURE 4.91 Prime mover operating speed (rad/s).

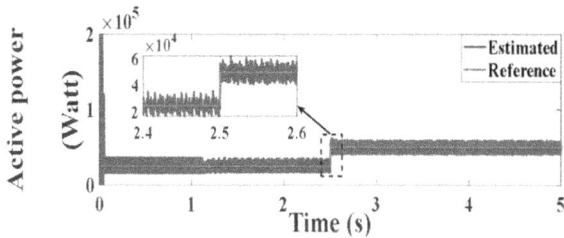

FIGURE 4.92 Active power under PVC (Watt).

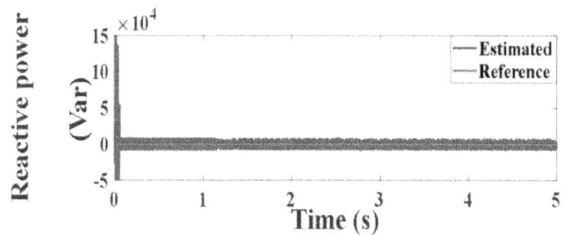

FIGURE 4.93 Reactive power under PVC (Var).

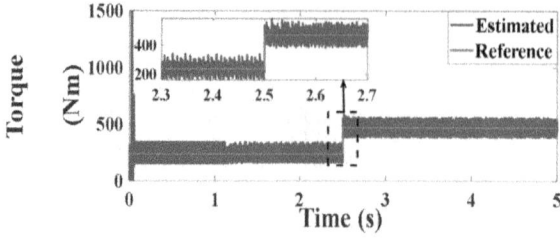

FIGURE 4.94 Torque developed with PVC (Nm).

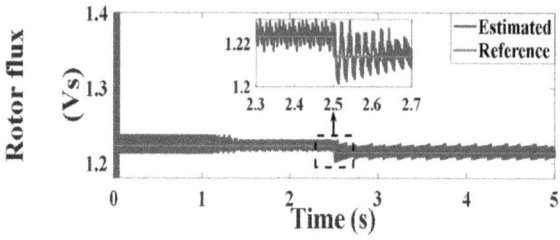

FIGURE 4.95 Rotor flux with PVC (Vs).

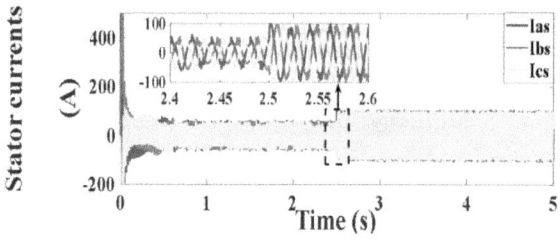

FIGURE 4.96 Stator currents under PVC (A).

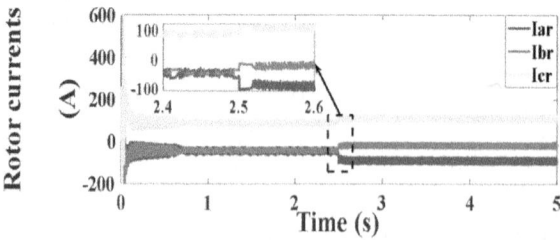

FIGURE 4.97 Rotor currents under PVC (A).

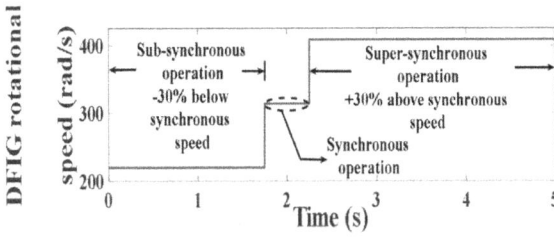

FIGURE 4.98 Prime mover operating speeds (rad/s).

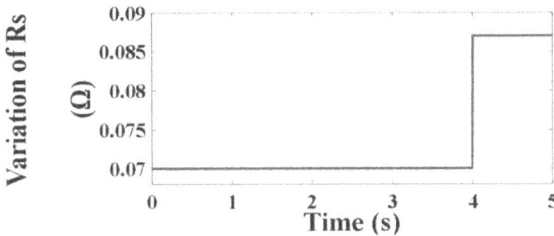

FIGURE 4.99 Variation of R_s under proposed PVC with variable operating speed (Ω).

4.2.2.5 Testing under Proposed PVC with Parameter Variation

In this section, a robustness test for the proposed PVC and another control scheme recently introduced in [82, 86–88] which adopts a similar operation mechanism is introduced. The purpose of this is to visualize the effectiveness of the proposed PVC in terms of system robustness.

4.2.2.5.1 Testing under Proposed PVC with 20% Variation of R_s

To visualize the performance of the designed PVC under parameter mismatch, the following tests were performed. For the variable speed operation shown in Figure 4.98, a mismatch in stator resistance of 20% was applied at time $t=4$ s (Figure 4.99). The obtained results reveal that the designed PVC succeeded in maintaining the proper operation under the resistance variation. This was confirmed by the maintained actual values of active and reactive powers, torque, and rotor flux (Figures 4.100–4.103) for variable speed operation. The same robust behavior is also maintained for fixed speed operation (Figure 4.104) under a resistance variation of 20% applied at time $t=2.5$ s (Figure 4.105). The obtained results for active and reactive powers (Figures 4.106 and 4.107) and for torque, and rotor flux (Figures 4.108 and 4.109) are confirming the robust dynamic behavior of the DFIG.

4.2.2.5.2 Testing under Proposed PVC with 20% Variation of R_r

The second parameter variation was made with the rotor resistance value with an increase of 20%. The values of active and reactive powers, torque, and rotor flux

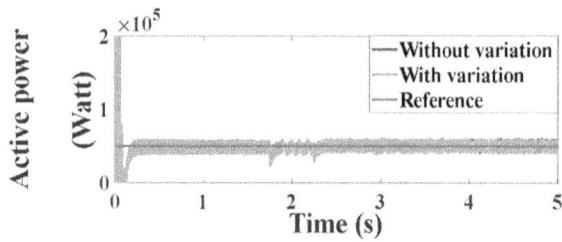

FIGURE 4.100   Active power under PVC (Watt).

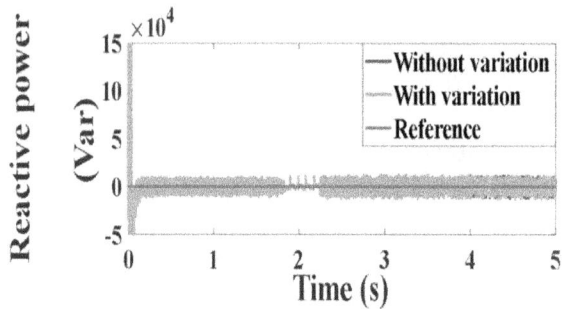

FIGURE 4.101   Reactive power under PVC (Var).

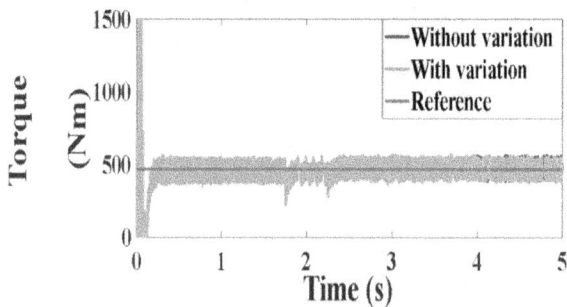

FIGURE 4.102   Torque developed with PVC (Nm).

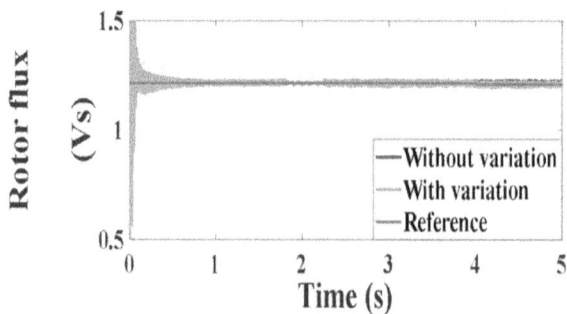

FIGURE 4.103   Rotor flux with PVC (Vs).

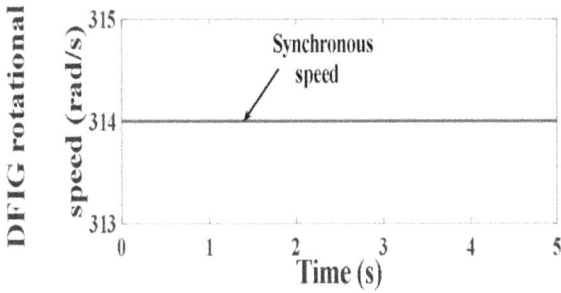

FIGURE 4.104   Prime mover operating speeds (rad/s).

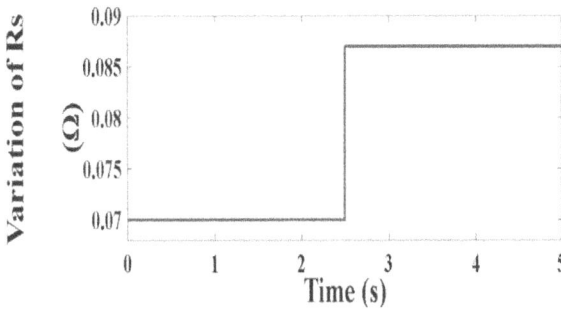

FIGURE 4.105   Variation of R_s under proposed PVC with constant operating speed (Ω).

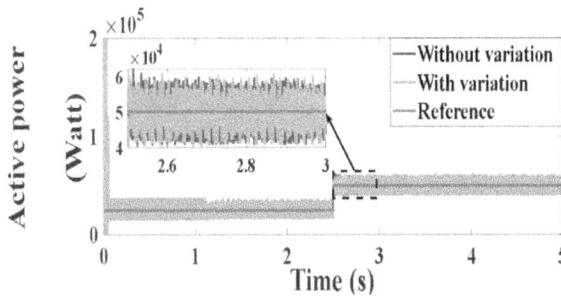

FIGURE 4.106   Active power under PVC (Watt).

for variable wind speed shown in Figure 4.110 with R_s variation (Figure 4.111) are presented in Figures 4.112–4.115, respectively. Similarly, for the fixed speed operation (Figure 4.116) with R_s variation (Figure 4.117), the results are illustrated for the same variables in Figures 4.118–4.121, respectively. The obtained results present an adequate performance under the resistance variation, which again confirms the robustness of the designed PVC scheme.

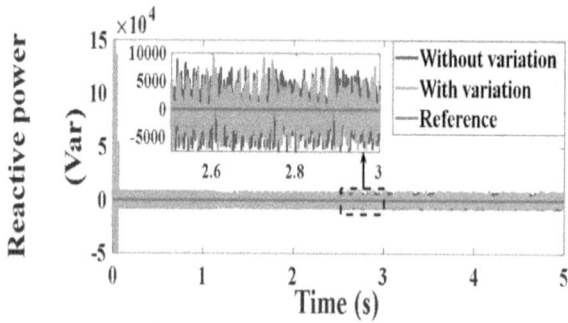

FIGURE 4.107   Reactive power under PVC (Var).

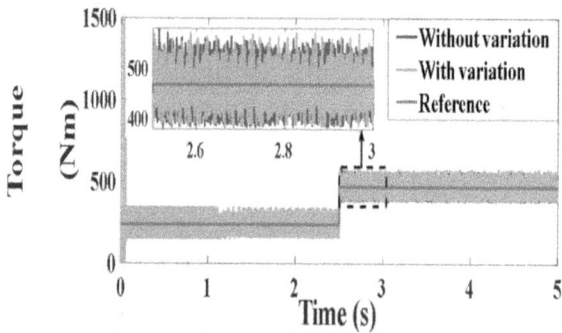

FIGURE 4.108   Torque developed with PVC (Nm).

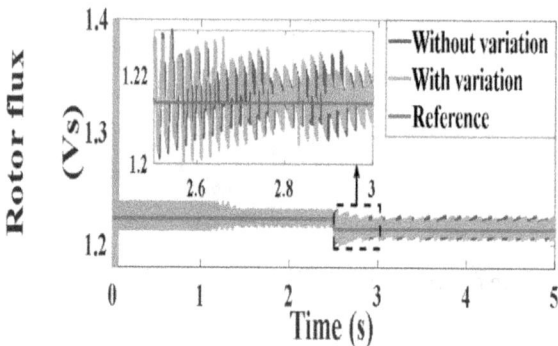

FIGURE 4.109   Rotor flux with PVC (Vs).

4.2.2.5.3   Testing under Proposed PVC with 15% Variation of L_s

The performance of the DFIG under PVC was also tested for the variable speed operation (Figure 4.122) while considering a mismatch of 15% in the stator inductance (L_s) value (Figure 4.123). The obtained results for the active and reactive powers,

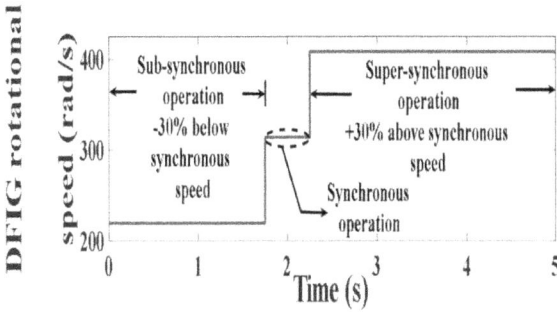

FIGURE 4.110 Prime mover operating speeds (rad/s).

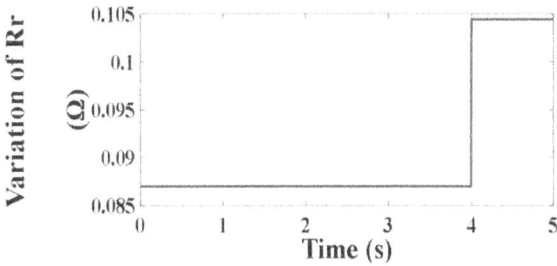

FIGURE 4.111 Variation of R_r (Ω) under proposed PVC with variable operating speed.

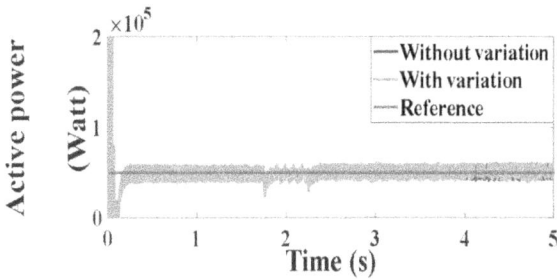

FIGURE 4.112 Active power under PVC (Watt).

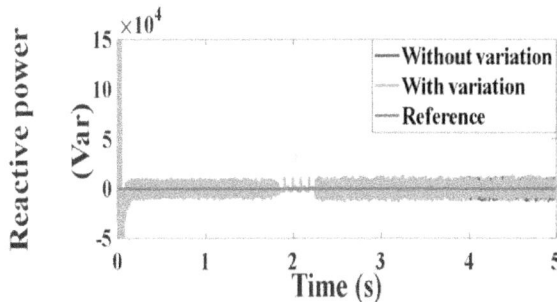

FIGURE 4.113 Reactive power under PVC (Var).

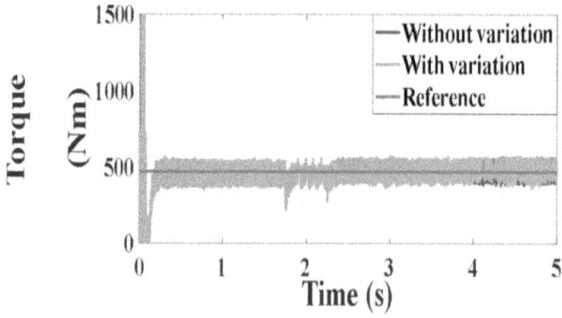

FIGURE 4.114 Torque developed with PVC (Nm).

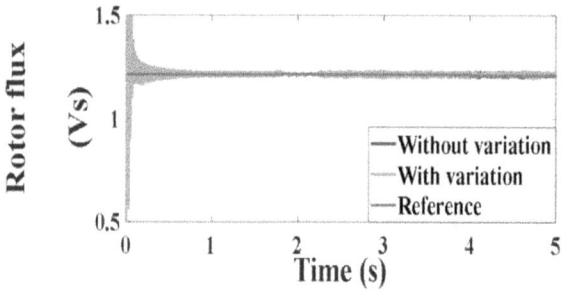

FIGURE 4.115 Rotor flux with PVC (Vs).

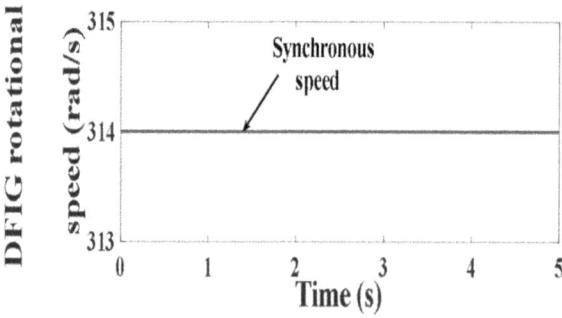

FIGURE 4.116 Prime mover operating speeds (rad/s).

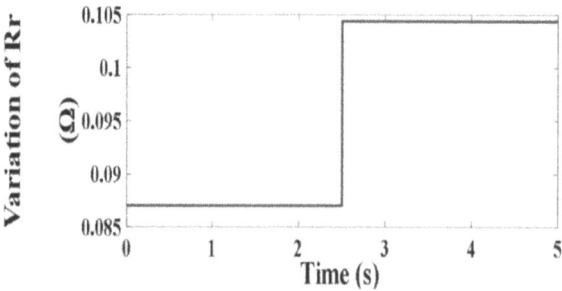

FIGURE 4.117 Variation of R_r (Ω).

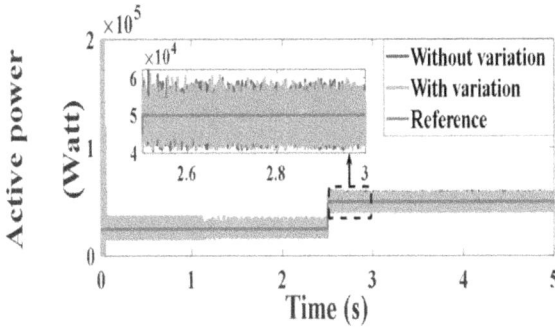

FIGURE 4.118   Active power under PVC (Watt).

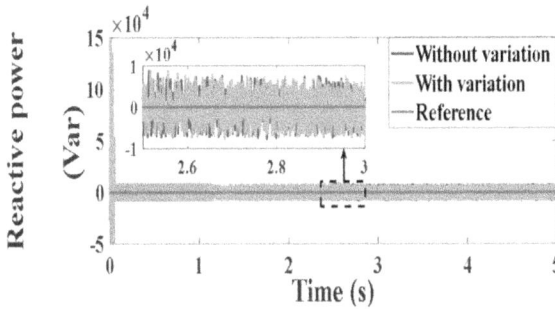

FIGURE 4.119   Reactive power under PVC (Var).

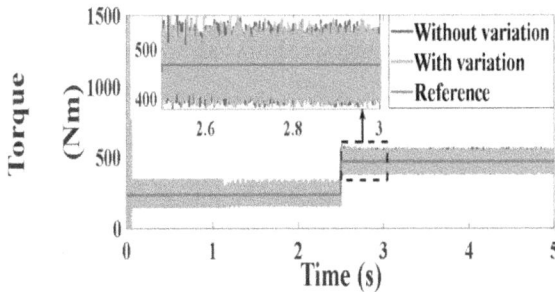

FIGURE 4.120   Torque developed with PVC (Nm).

torque and rotor flux are presented in Figures 4.124–4.127. A similar test is applied under the fixed speed operation (Figure 4.128) with a L_s variation (Figure 4.129), and the results are presented in Figures 4.130–4.133, respectively. The obtained results reveal that the DFIG maintains an appropriate operation even under the inductance variation.

4.2.2.5.4   Testing under Proposed PVC with 15% Variation of L_r

The performance of the DFIG under PVC was also tested for the variable speed operation (Figure 4.134) while considering a mismatch of 15% in the stator inductance

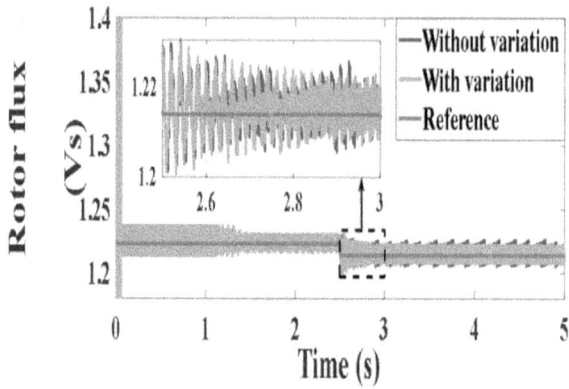

FIGURE 4.121   Rotor flux with PVC (Vs).

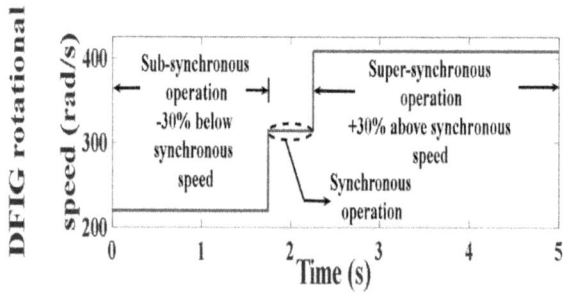

FIGURE 4.122   Prime mover operating speeds (rad/s).

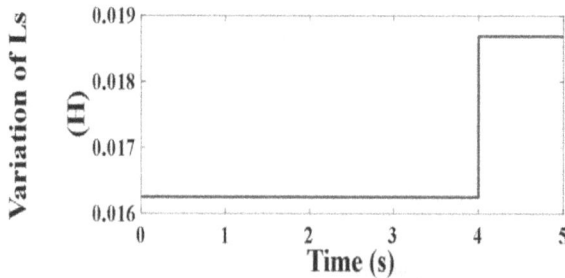

FIGURE 4.123   Variation of L_s (H) under proposed PVC with variable operating speed.

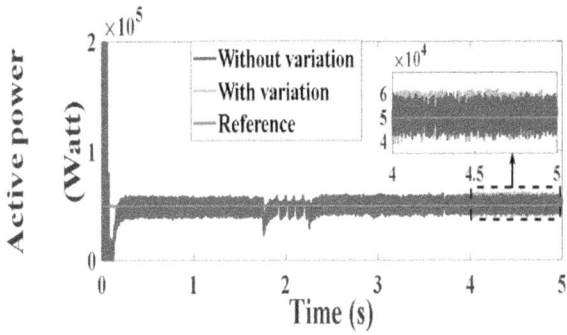

FIGURE 4.124   Active power under PVC (Watt).

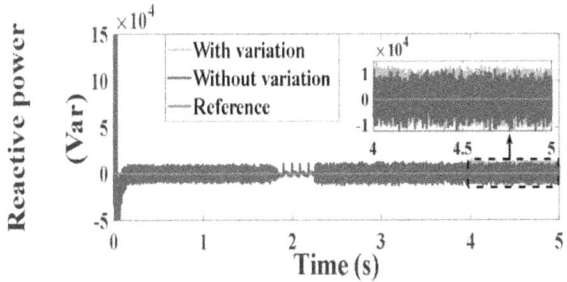

FIGURE 4.125   Reactive power under PVC (Var).

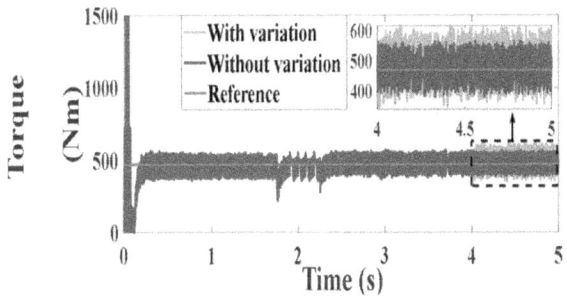

FIGURE 4.126   Torque developed with PVC (Nm).

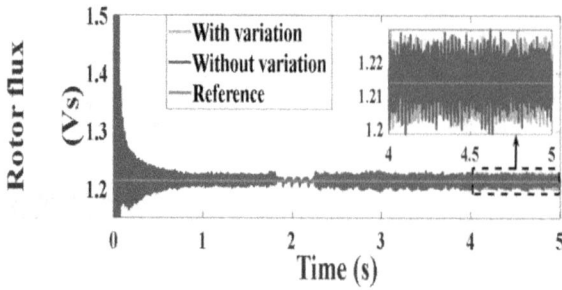

FIGURE 4.127   Rotor flux with PVC (Vs).

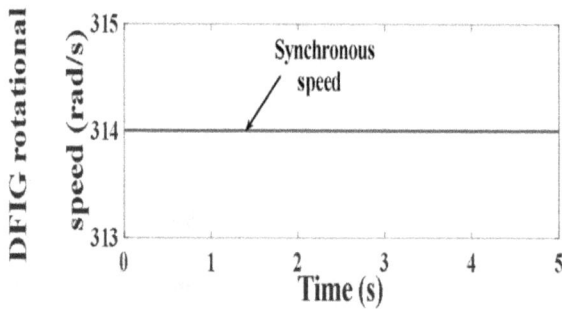

FIGURE 4.128   Prime mover operating speeds (rad/s).

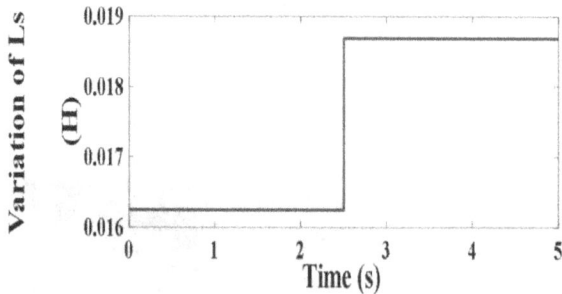

FIGURE 4.129   Variation of L_s (H) under proposed PVC with variable operating speed.

(L_r) value (Figure 4.135). The obtained results for the active and reactive powers, torque and rotor flux are presented in Figures 4.136–4.139. A similar test is applied under the fixed speed operation (Figure 4.140) with a L_r variation (Figure 4.141), and the results are presented in Figures 4.142–4.145, respectively. The captured figures confirm the ability of the designed PVC to keep the actual values within their pre-defined references.

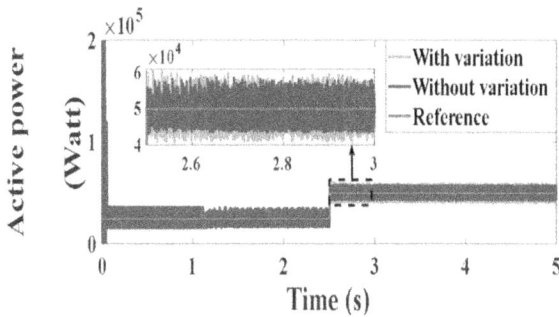

FIGURE 4.130   Active power under PVC (Watt).

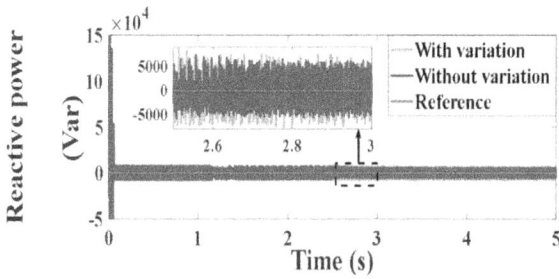

FIGURE 4.131   Reactive power under PVC (Var).

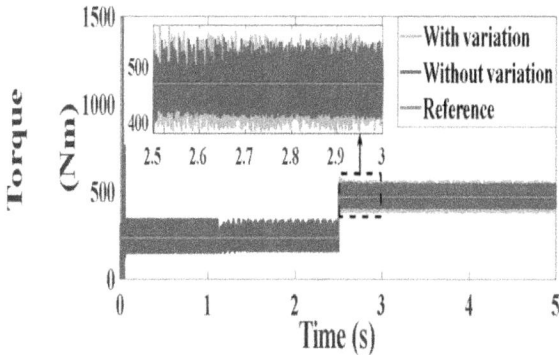

FIGURE 4.132   Torque developed with PVC (Nm).

4.2.2.5.5   Testing under Proposed PVC with 15% Variation of L_m

The last performance test of the DFIG for the variable speed operation (Figure 4.146) while considering a mismatch of 15% in the magnetizing inductance (L_m) value (Figure 4.147). The obtained results for the active and reactive powers, torque and rotor flux are presented in Figures 4.148–4.151. A similar test is applied under the

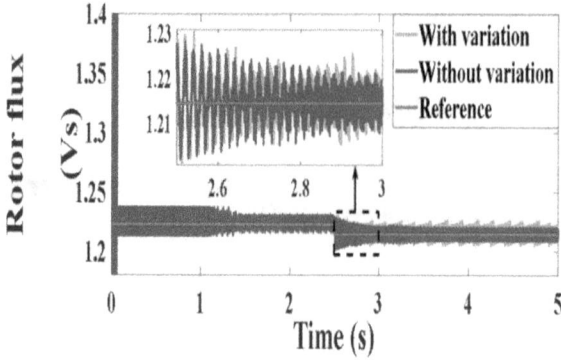

FIGURE 4.133   Rotor flux with PVC (Vs).

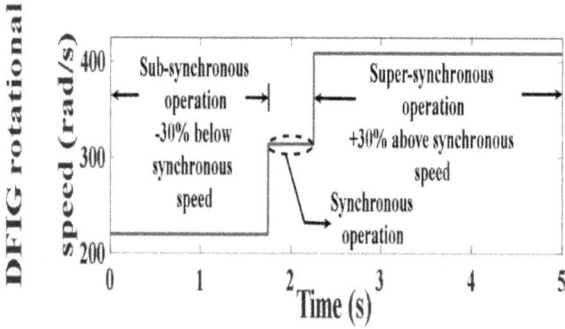

FIGURE 4.134   Prime mover operating speeds (rad/s).

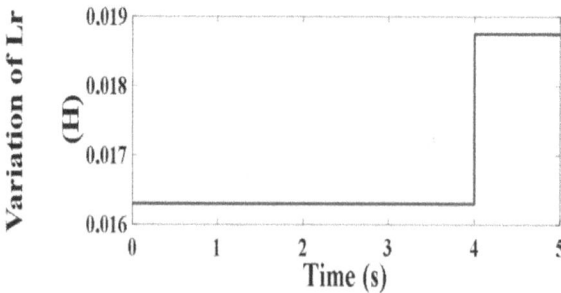

FIGURE 4.135   Variation of L_r (H) under proposed PVC with variable operating speed.

fixed speed operation (Figure 4.152) with a L_m variation (Figure 4.153), and the results are presented in Figures 4.154–4.157, respectively. The results reveal that the actual variables still stayed within their relevant reference ranges, which validates the effectiveness of the designed PVC.

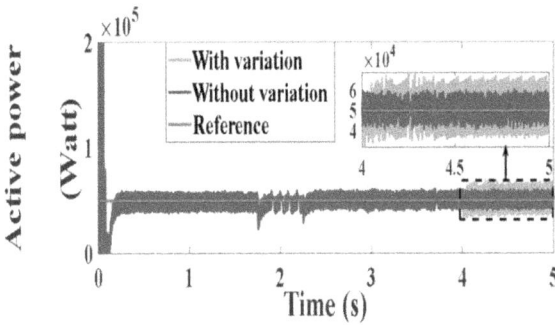

FIGURE 4.136   Active power under PVC (Watt).

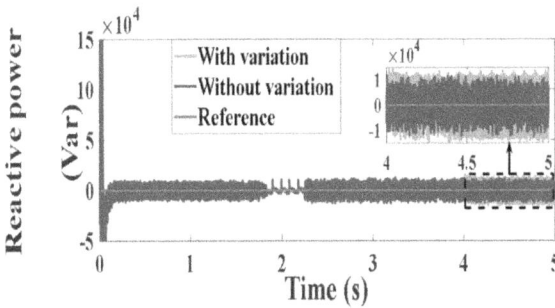

FIGURE 4.137   Reactive power under PVC (Var).

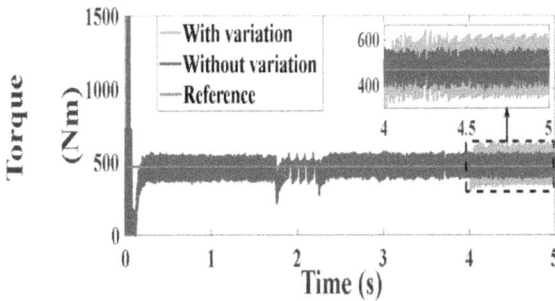

FIGURE 4.138   Torque developed with PVC (Nm).

4.2.2.6   Testing Using Deadbeat-Based Predictive Control (DBPC) Scheme with Parameter Variation

4.2.2.6.1   Testing the DBPC Scheme with 20% Variation of R_s

To visualize the performance of the DFIG under the PVC technique, which uses the deadbeat principle [82, 86–88] for generating the reference voltages

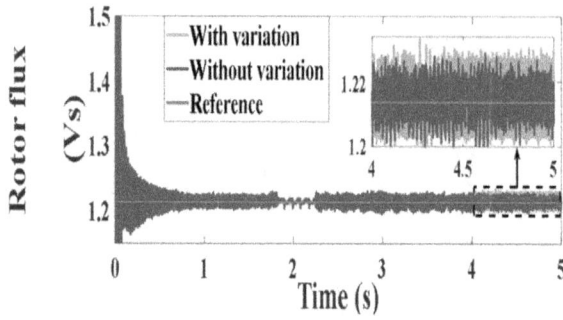

FIGURE 4.139 Rotor flux with PVC (Vs).

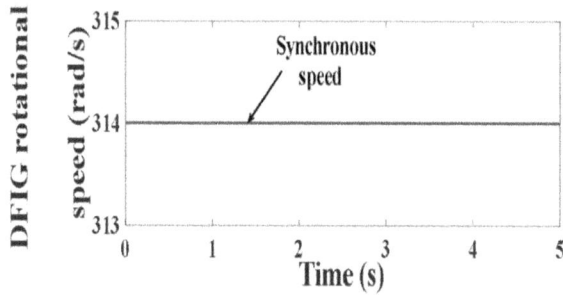

FIGURE 4.140 Prime mover operating speeds (rad/s).

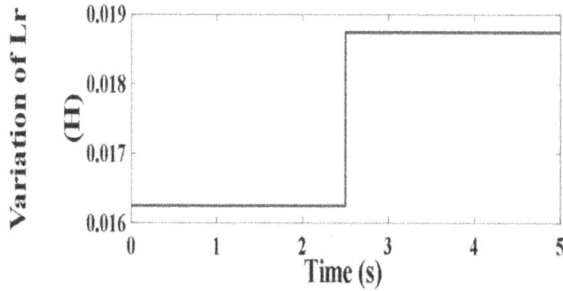

FIGURE 4.141 Variation of L_r (H).

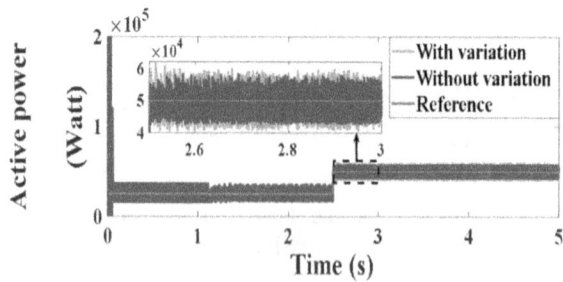

FIGURE 4.142 Active power under PVC (Watt).

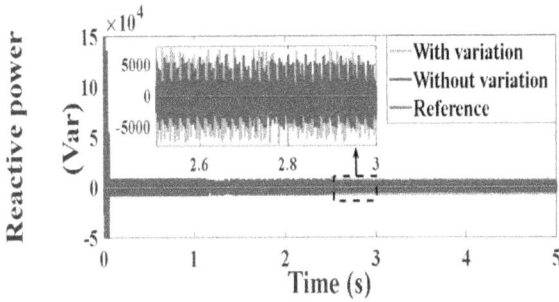

FIGURE 4.143 Reactive power under PVC (Var).

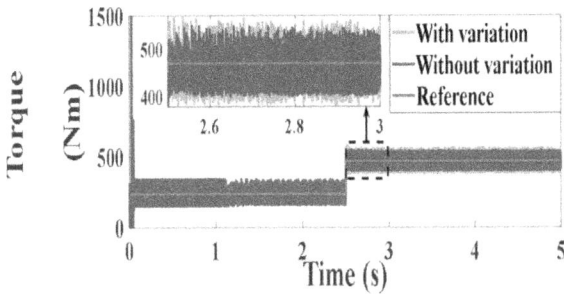

FIGURE 4.144 Torque developed with PVC (Nm).

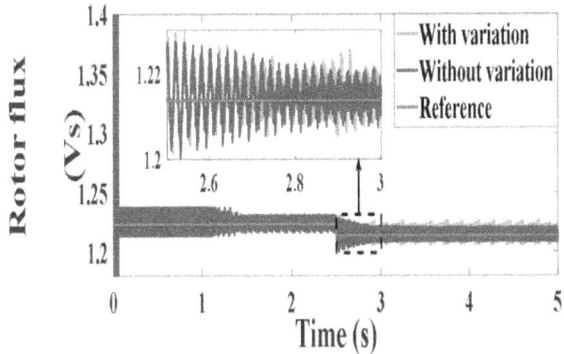

FIGURE 4.145 Rotor flux with PVC (Vs).

under parameter mismatch, the following tests were performed. This section presents the dynamic performance of the DFIG for variable speed operation (Figure 4.158) when a mismatch in stator resistance of 20% was applied at time $t = 4$ s (Figure 4.159). The obtained results reveal that this scheme of PVC failed in maintaining the proper operation under the resistance variation. This fact can be confirmed by following up the deviation of the actual values of active and reactive powers, torque, and rotor flux (Figures 4.160–4.163, respectively). A similar test under fixed speed operation (Figure 4.164) is applied with a resistance variation at time $t = 2.5$ s (Figure 4.165). The obtained results in Figures 4.166–4.169 reveal

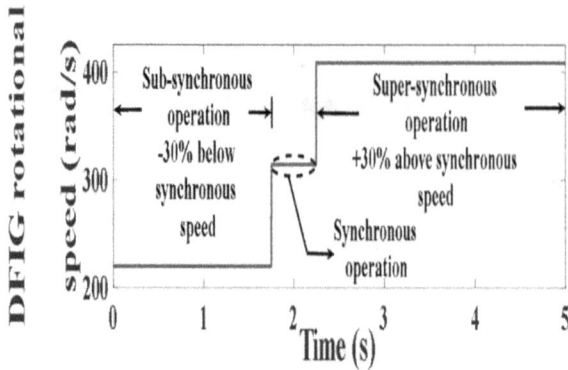

FIGURE 4.146 Prime mover operating speeds (rad/s).

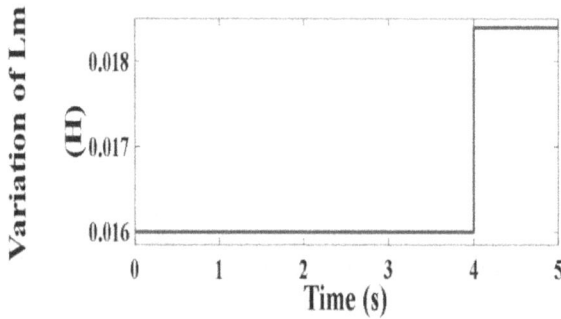

FIGURE 4.147 Variation of L_m (H) under proposed PVC with variable operating speed.

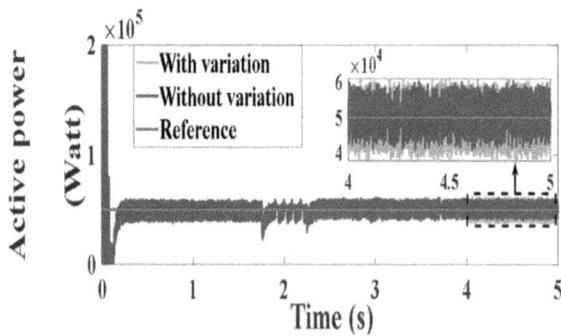

FIGURE 4.148 Active power under PVC (Watt).

the obvious deviation of the actual values from their references. As a result, it is confirmed that the validity and robustness of our designed PVC against parameter change compared with this scheme is ensured.

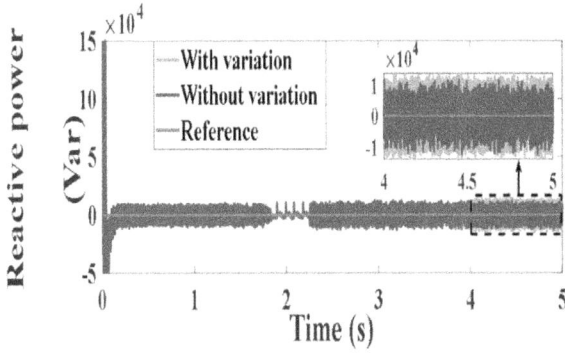

FIGURE 4.149   Reactive power under PVC (Var).

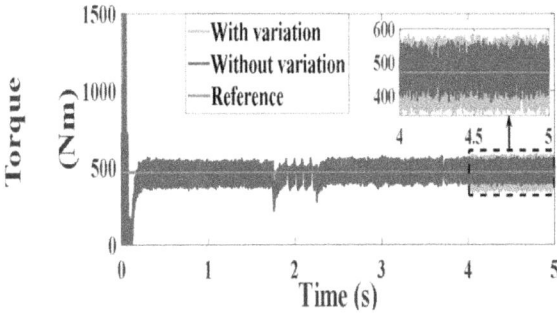

FIGURE 4.150   Torque developed with PVC (Nm).

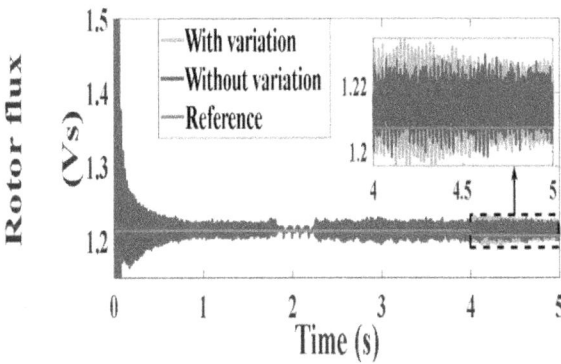

FIGURE 4.151   Rotor flux with PVC (Vs).

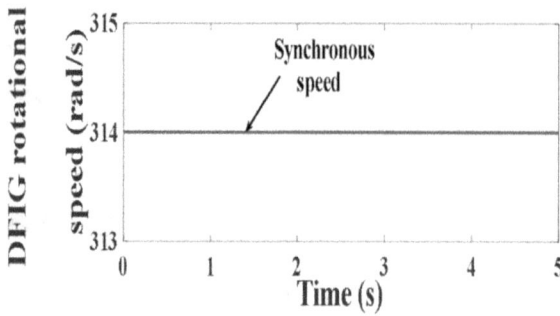

FIGURE 4.152   Prime mover operating speeds (rad/s).

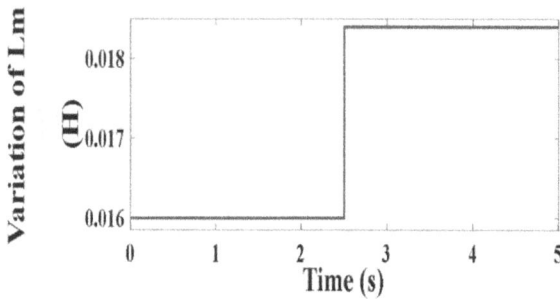

FIGURE 4.153   Variation of L_m (H).

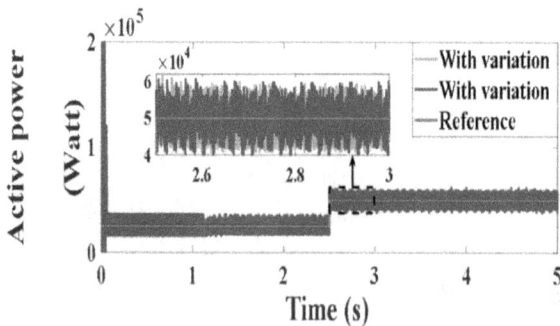

FIGURE 4.154   Active power under PVC (Watt).

4.2.2.6.2   Testing the DBPC Scheme with 20% Variation of R_r

The dynamic performance of the DFIG for variable speed operation (Figure 4.170) when a mismatch in rotor resistance of 20% was applied (Figure 4.171). The obtained results reveal that DBPC scheme failed in maintaining the proper operation under the resistance variation. This fact can be confirmed by following up the deviation of

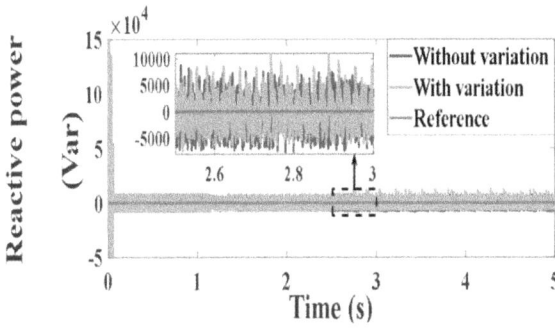

FIGURE 4.155   Reactive power under PVC (Var).

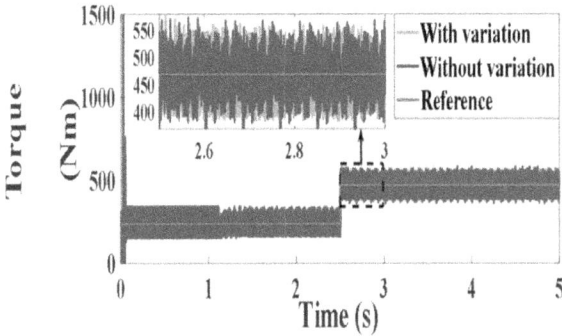

FIGURE 4.156   Torque developed with PVC (Nm).

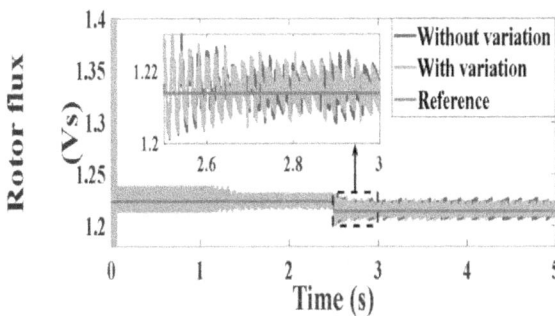

FIGURE 4.157   Rotor flux with PVC (Vs).

the actual values of active and reactive powers, torque, and rotor flux (Figures 4.172–4.175, respectively). A similar test under fixed speed operation (Figure 4.176) is applied with a resistance variation at time $t = 2.5\,\text{s}$ (Figure 4.177). The obtained results in Figures 4.178–4.181 reveal the obvious deviation of the actual values from their references.

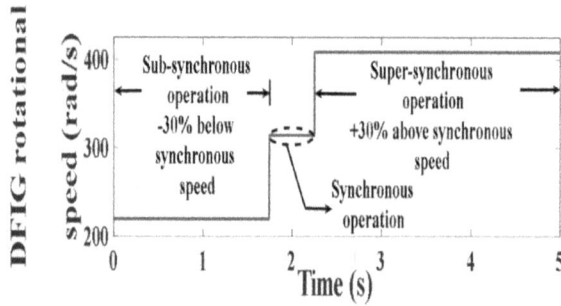

FIGURE 4.158   Prime mover operating speeds (rad/s).

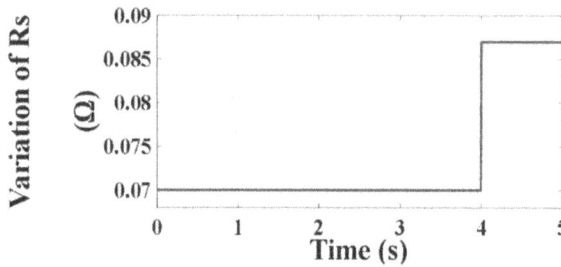

FIGURE 4.159   Variation of R_s (Ω).

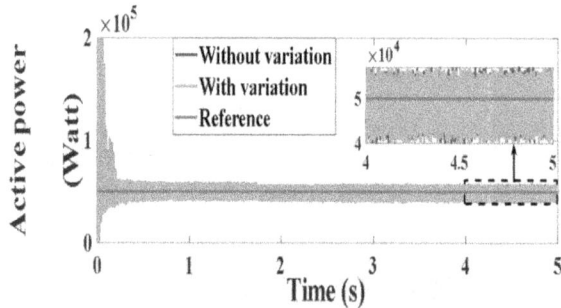

FIGURE 4.160   Active power under PVC (Watt).

4.2.2.6.3   Testing the DBPC Scheme with 15% Variation of L_s

The performance of the DFIG under the DBPC scheme was also tested while considering a mismatch of 15% in the stator inductance (L_s) value. For the variable speed operation (Figure 4.182) when a mismatch in L_s of 15% was applied (Figure 4.183), the obtained results reveal that DBPC scheme failed in maintaining the proper operation under the inductance variation. This fact can be confirmed by following up the deviation of the actual values of active and reactive powers, torque and rotor flux (Figures 4.184–4.187, respectively). A similar test under fixed speed operation

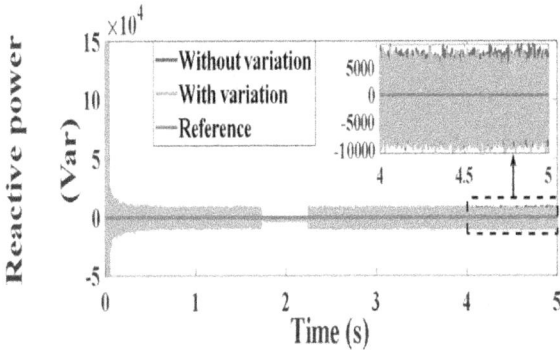

FIGURE 4.161 Reactive power under PVC (Var).

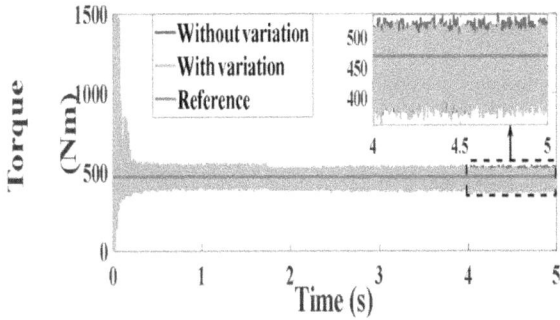

FIGURE 4.162 Torque developed with PVC (Nm).

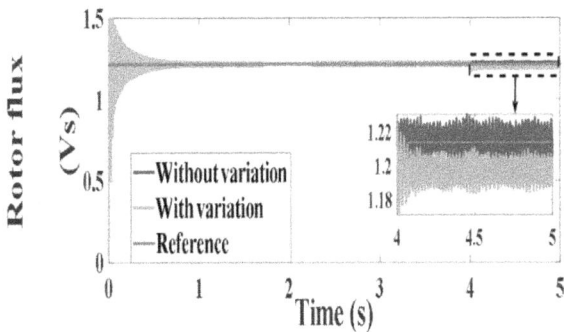

FIGURE 4.163 Rotor flux with PVC (Vs).

(Figure 4.188) is applied with an inductance variation at time $t = 2.5$ s (Figure 4.189). The Computation Results in Figures 4.190–4.193 prove the effectiveness and robustness of our designed PVC against parameter change compared with the DBPC scheme.

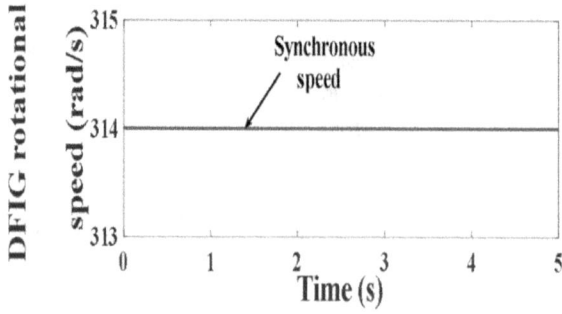

FIGURE 4.164 Prime mover operating speeds (rad/s).

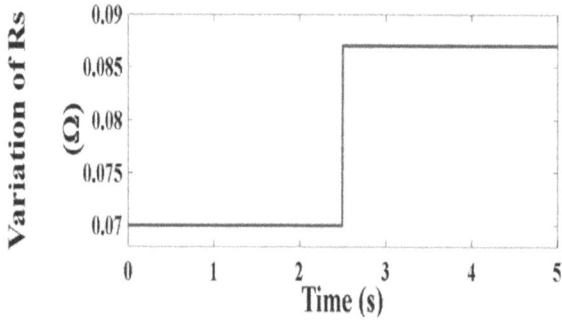

FIGURE 4.165 Variation of R_s (Ω).

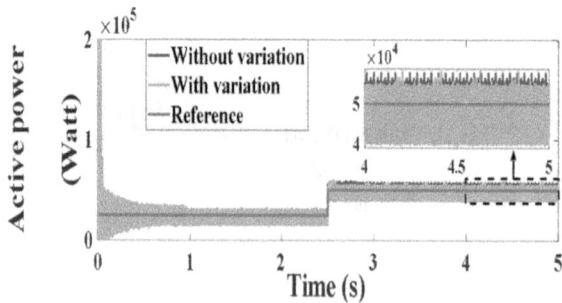

FIGURE 4.166 Active power under PVC (Watt).

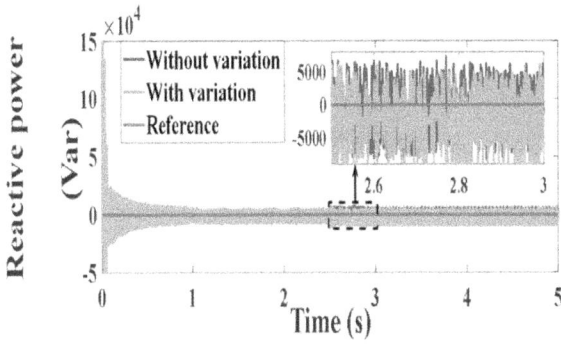

FIGURE 4.167   Reactive power under PVC (Var).

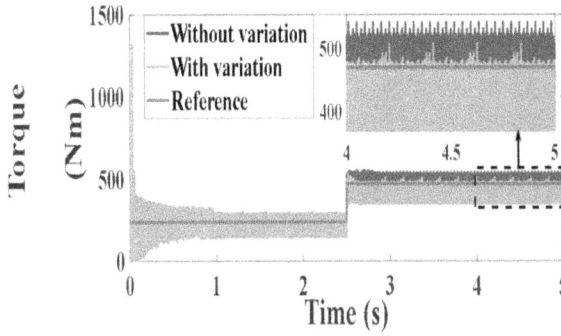

FIGURE 4.168   Torque developed under PVC (Nm).

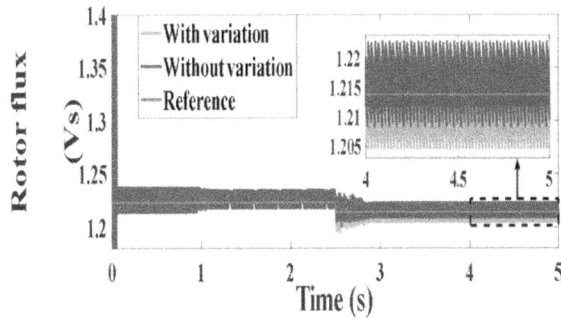

FIGURE 4.169   Rotor flux under PVC (Vs).

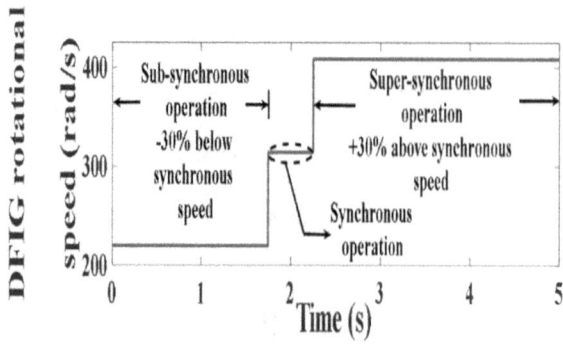

FIGURE 4.170   Prime mover operating speeds (rad/s).

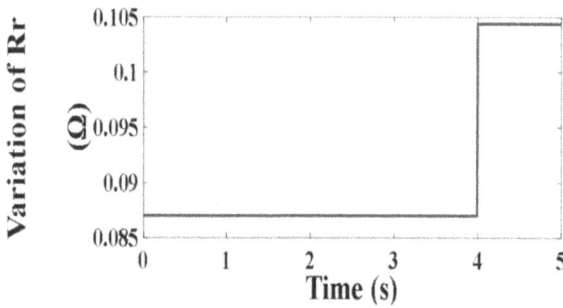

FIGURE 4.171   Variation of R_r (Ω)

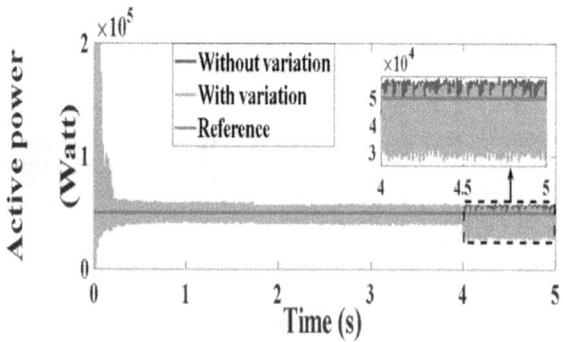

FIGURE 4.172   Active power under PVC (Watt).

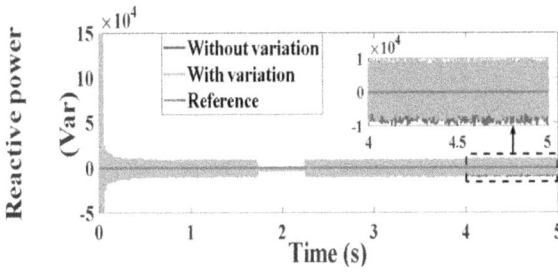

FIGURE 4.173   Reactive power under PVC (Var).

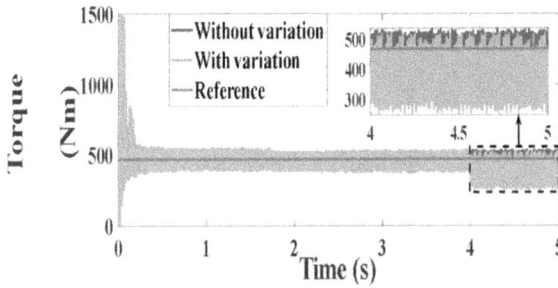

FIGURE 4.174   Torque developed with PVC (Nm).

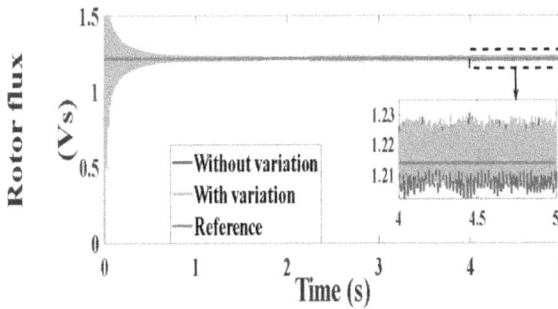

FIGURE 4.175   Rotor flux with PVC (Vs).

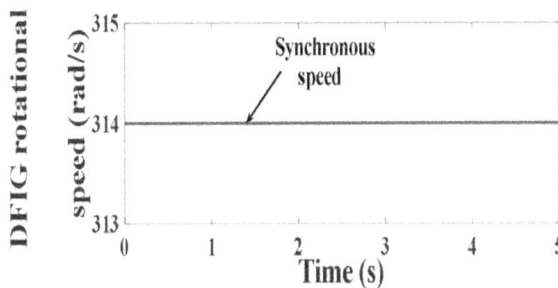

FIGURE 4.176   Prime mover operating speeds (rad/s).

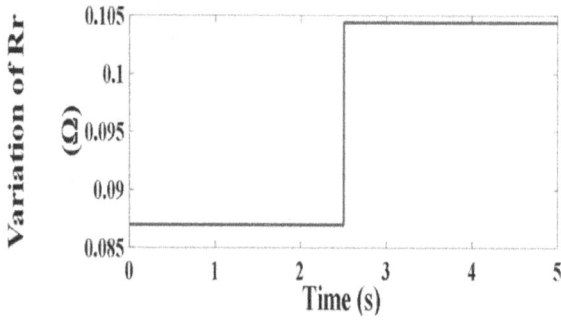

FIGURE 4.177   Variation of R_r (Ω).

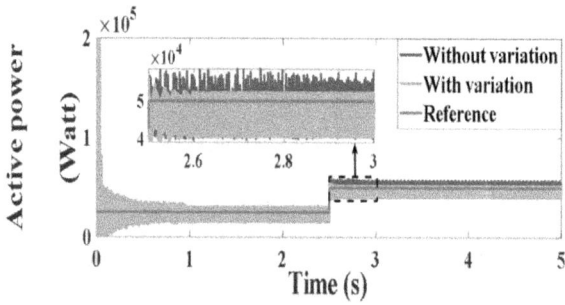

FIGURE 4.178   Active power under PVC (Watt).

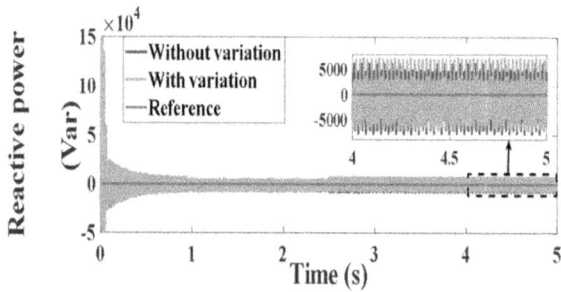

FIGURE 4.179   Reactive power under PVC (Var).

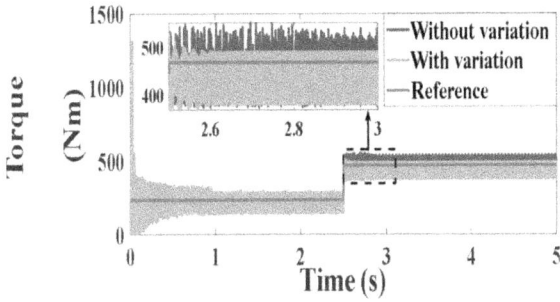

FIGURE 4.180   Torque developed with PVC (Nm).

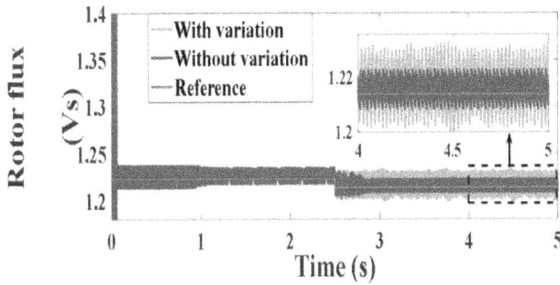

FIGURE 4.181   Rotor flux with PVC (Vs).

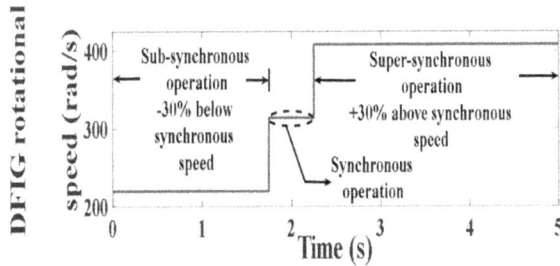

FIGURE 4.182   Prime mover operating speeds (rad/s).

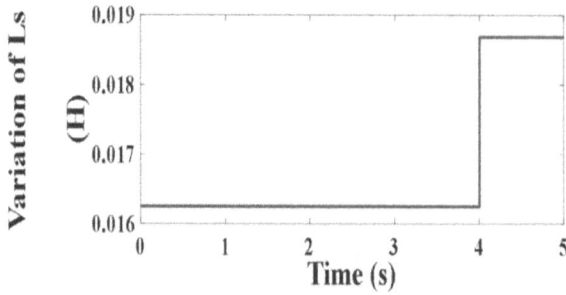

FIGURE 4.183   Variation of L_s (H).

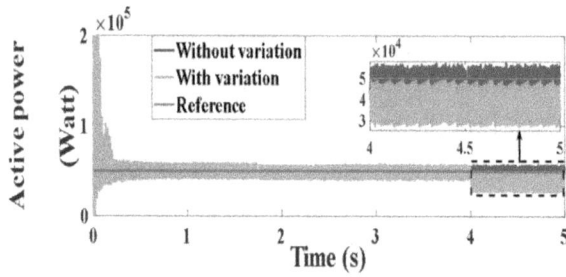

FIGURE 4.184   Active power under PVC (Watt).

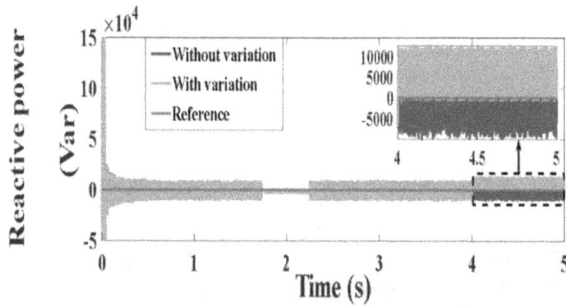

FIGURE 4.185   Reactive power under PVC (Var).

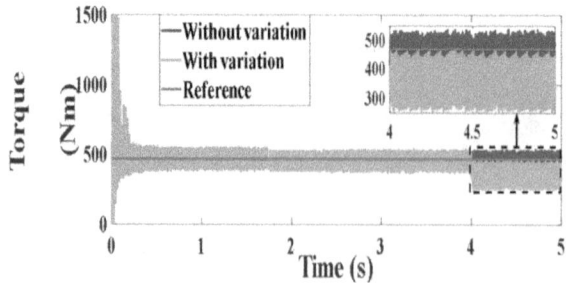

FIGURE 4.186   Torque developed with PVC (Nm).

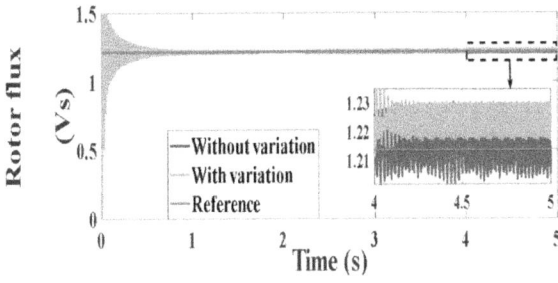

FIGURE 4.187   Rotor flux with PVC (Vs).

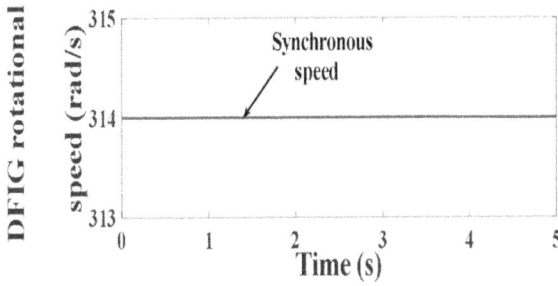

FIGURE 4.188   Prime mover operating speeds (rad/s).

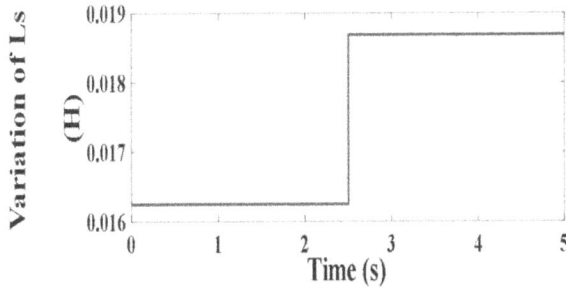

FIGURE 4.189   Variation of L_s (H).

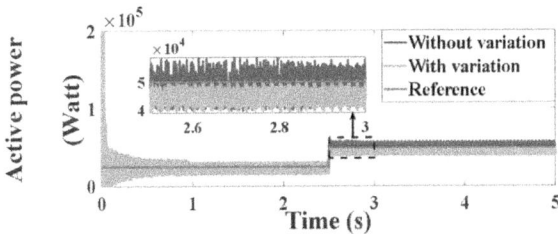

FIGURE 4.190   Active power under PVC (Watt).

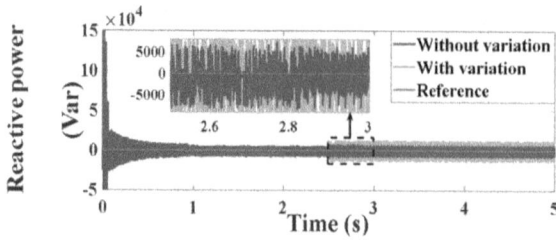

FIGURE 4.191   Reactive power under PVC (Var).

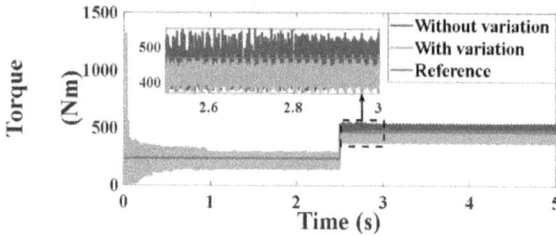

FIGURE 4.192   Torque developed with PVC (Nm).

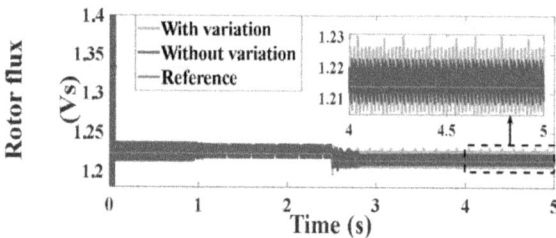

FIGURE 4.193   Rotor flux with PVC (Vs).

4.2.2.6.4 Testing the DBPC Scheme with 15% Variation of L_r

For the variable speed operation (Figure 4.194), a mismatch in L_r of 15% was applied (Figure 5.195). The obtained results reveal that DBPC scheme failed in maintaining the proper operation under the inductance variation. This fact is illustrated via checking the deviation of the actual values of active and reactive powers, torque, and rotor flux (Figures 4.196–4.199, respectively). A similar test under fixed speed operation (Figure 4.200) is applied with a L_r variation at time $t = 2.5\,\mathrm{s}$ (Figure 4.201). The results in Figures 4.202–4.205 reveal the deficiency of the DBPC in handling the system uncertainties.

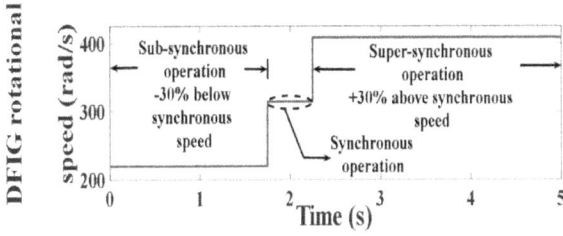

FIGURE 4.194   Prime mover operating speeds (rad/s).

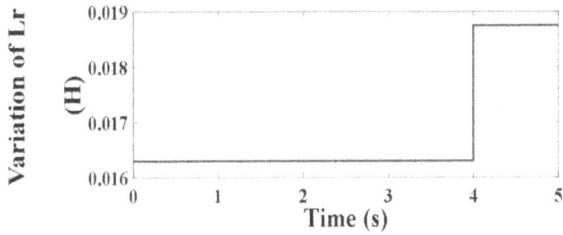

FIGURE 4.195   Variation of L_r (H).

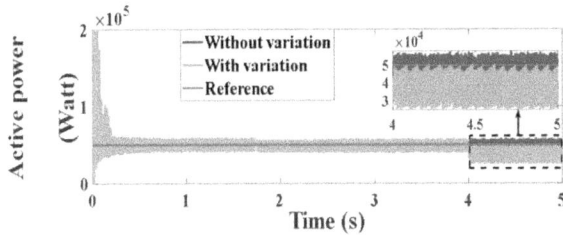

FIGURE 4.196   Active power under PVC (Watt).

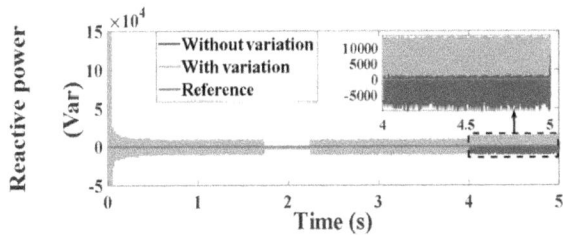

FIGURE 4.197   Reactive power under PVC (Var).

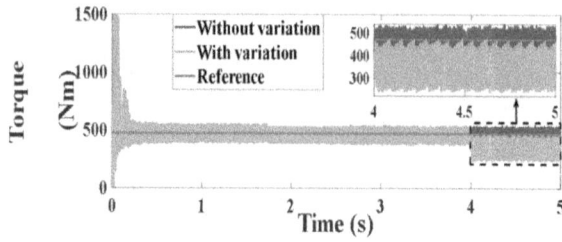

FIGURE 4.198   Torque developed with PVC (Nm).

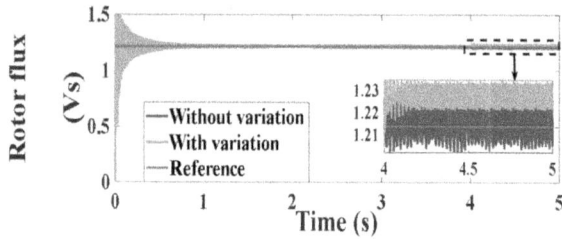

FIGURE 4.199   Rotor flux with PVC (Vs).

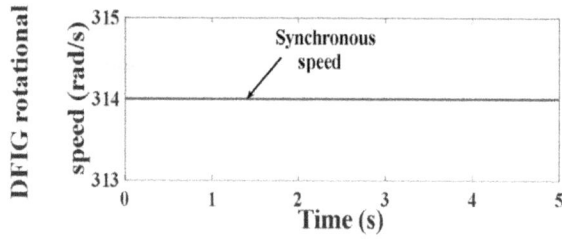

FIGURE 4.200   Prime mover operating speeds (rad/s).

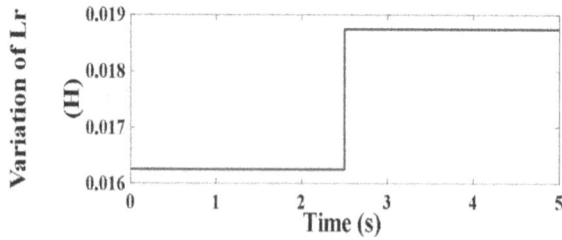

FIGURE 4.201   Variation of L_r (H).

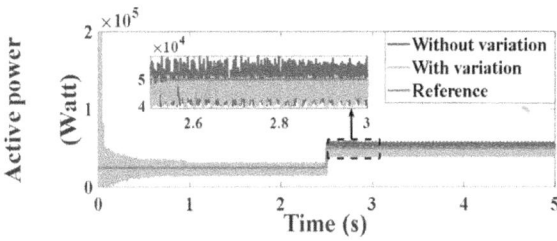

FIGURE 4.202 Active power under PVC (Watt).

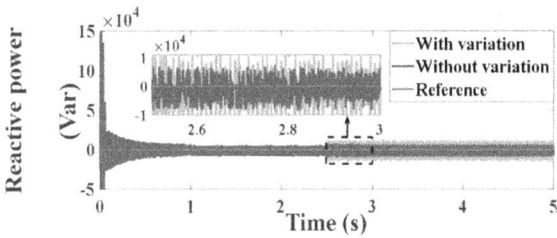

FIGURE 4.203 Reactive power under PVC (Var).

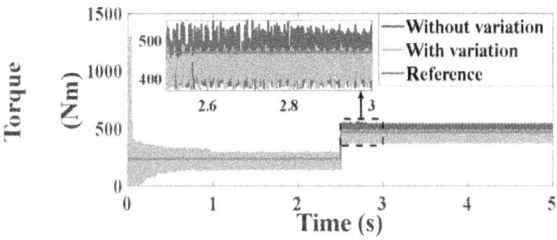

FIGURE 4.204 Torque developed with PVC (Nm).

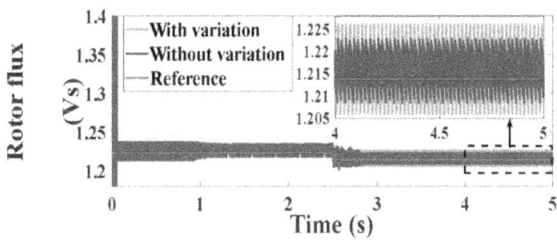

FIGURE 4.205 Rotor flux with PVC (Vs).

4.2.2.6.5 Testing the DBPC Scheme with 15% Variation of L_m

The last performance test of the DFIG for the variable speed operation (Figure 4.206) while considering a mismatch of 15% in the magnetizing inductance (L_m) value (Figure 4.207). The obtained results for the active and reactive powers, torque and rotor flux are presented in Figures 4.208–4.211. A similar test is applied under the fixed speed operation (Figure 4.212) with a L_m variation (Figure 4.213), and the results are presented in Figures 4.214–4.217, respectively. The results reveal that the actual variables still deviated from their relevant reference ranges under the mutual inductance variation, which validates the robustness and superiority of our proposed PVC.

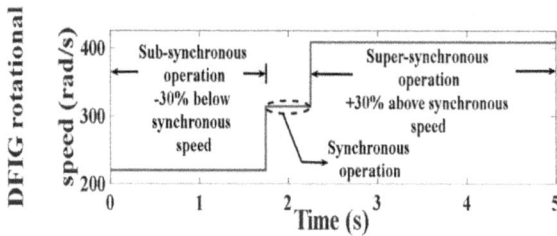

FIGURE 4.206   Prime mover operating speeds (rad/s).

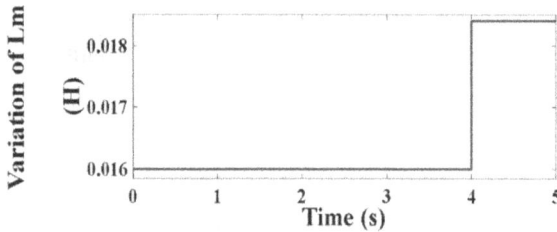

FIGURE 4.207   Variation of L_m (H).

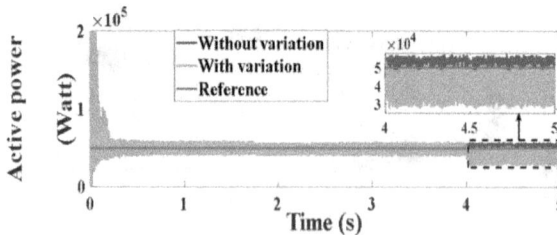

FIGURE 4.208   Active power under PVC (Watt).

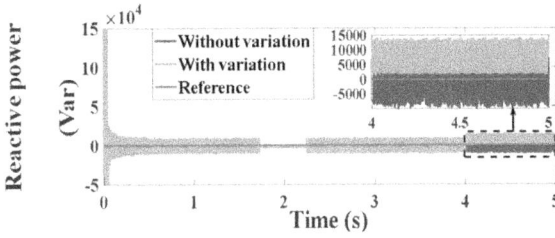

FIGURE 4.209   Reactive power under PVC (Var).

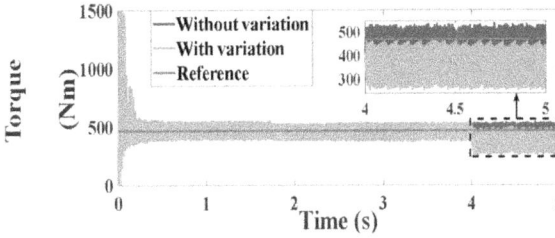

FIGURE 4.210   Torque developed with PVC (Nm).

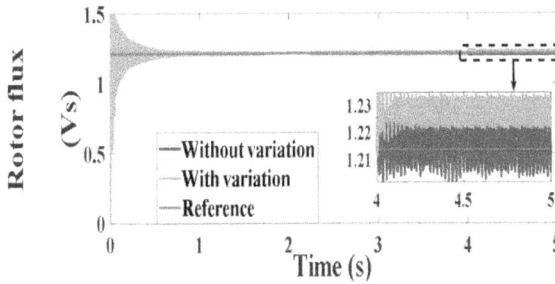

FIGURE 4.211   Rotor flux with PVC (Vs).

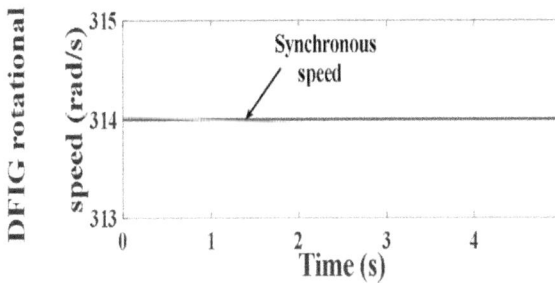

FIGURE 4.212   Prime mover operating speeds (rad/s).

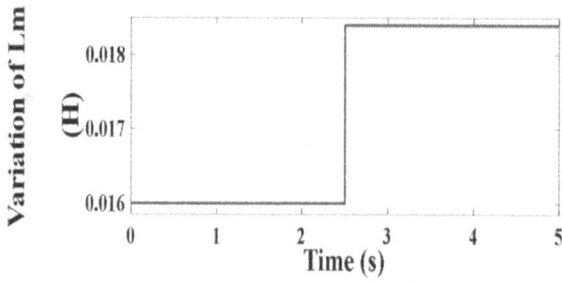

FIGURE 4.213 Variation of L_m (H).

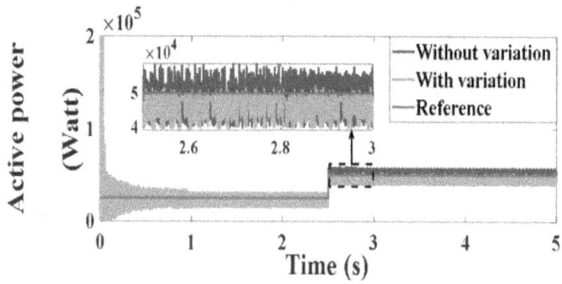

FIGURE 4.214 Active power under PVC (Watt).

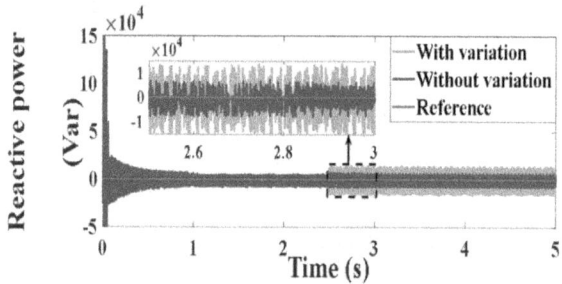

FIGURE 4.215 Reactive power under PVC (Var).

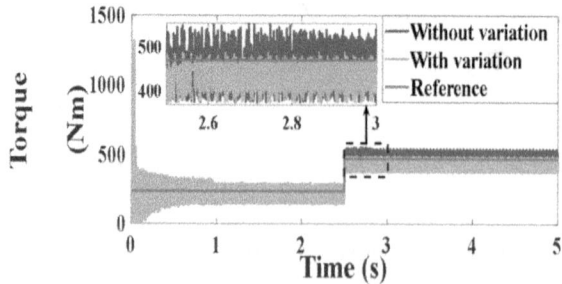

FIGURE 4.216 Torque developed with PVC (Nm).

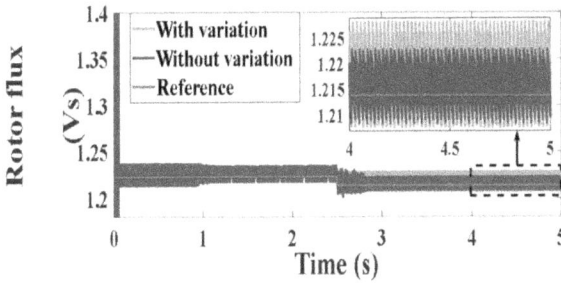

FIGURE 4.217   Rotor flux with PVC (Vs).

4.2.2.7   Comparison Study

Finally, a comparison was performed so that the dynamic performance of the DFIG can be studied under the four control techniques to enable the selection of the most appropriate technique to be used with the DFIG. The control effectiveness was measured in terms of ripples content, time of dynamic response, computational burden, and THD. The results are shown in Figures 4.218–4.222, which outline the active power, reactive power, developed torque, and rotor flux magnitude for the variable speed operation, and illustrated in Figures 4.223–4.227 the fixed speed operation, respectively. Table 4.4 and Figure 4.228 introduce a comparison of the response time for each technique to clarify, with the aid of the results, which technique has a shorter response time and consequently possesses the fastest dynamic response. As noted in Table 4.4, the actual values under the proposed PVC technique show a shorter response time in comparison with the other control approaches, as the controlled variables in PVC are the voltage vectors, which are the nearest electrical value applied to the machine windings. Table 4.5 and Figure 4.229 present a comparison of the ripples content in the four principles and prove, with the help of the results, that the ripples content of the PVC technique is lower than that of the MPCC and MPDTC approaches. Furthermore, the formulated approach presents a lower computational burden, as it has lower number of performed commutations compared with that of MPCC and MPDTC strategies, as shown in Table 4.6 and Figure 4.230.

Figures 4.231–4.248, present the spectrums of the stator current phases under the MPCC, MPDTC and designed PVC, the proposed controller presents a lower value of THD compared with that of MPCC and MPDTC as noticed from comparing the spectrums. This fact can be approved by the numerical values of Table 4.7 and the chart of Figure 4.249.

Eventually, it can thus be concluded that the proposed PVC technique is the most appropriate control scheme to be used with the DFIG for the following reasons: it eliminates system complexity; its dynamic response is the fastest compared with that of the MPCC, MPDTC, and SVOC principles; its ripples content is lower than that of the MPCC and MPDTC approaches; it has lower number of executed commutations compared with that of MPCC and MPDTC strategies; and moreover, it has the lowest value of THD.

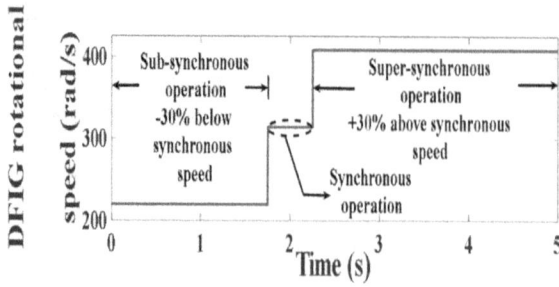

FIGURE 4.218   Prime mover operating speeds (rad/s).

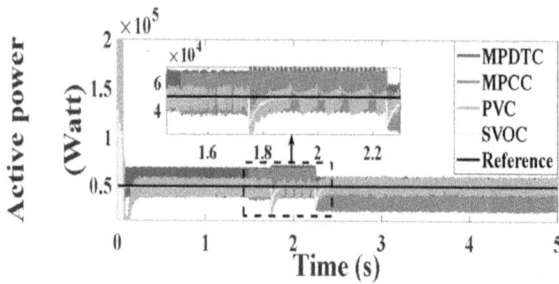

FIGURE 4.219   Comparison between active powers of four techniques (Watt).

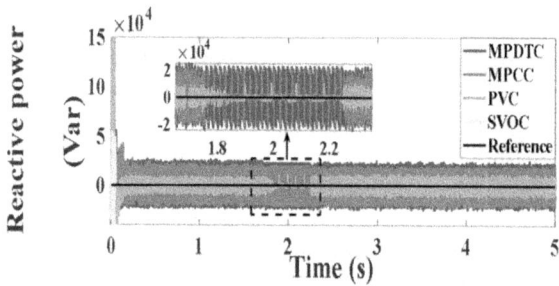

FIGURE 4.220   Comparison between reactive powers of four techniques with variable operating speed (Var).

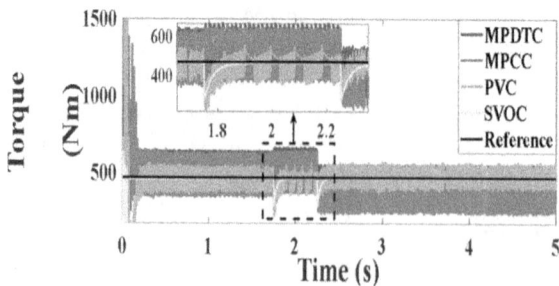

FIGURE 4.221   Comparison between torques of four techniques (Nm) with variable operating speed.

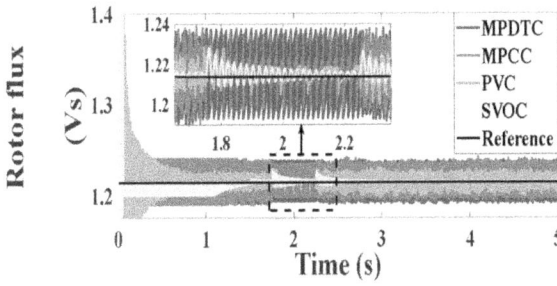

FIGURE 4.222   Comparison between rotor fluxes of four techniques (Vs) with variable operating speed.

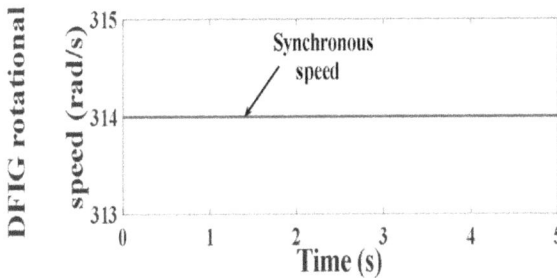

FIGURE 4.223   Prime mover operating speed (rad/s).

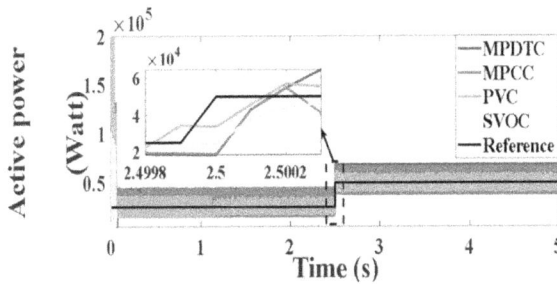

FIGURE 4.224   Comparison between active powers of four techniques (Watt).

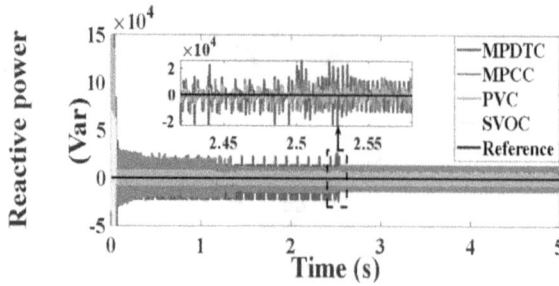

FIGURE 4.225 Comparison between reactive powers of four techniques (Var).

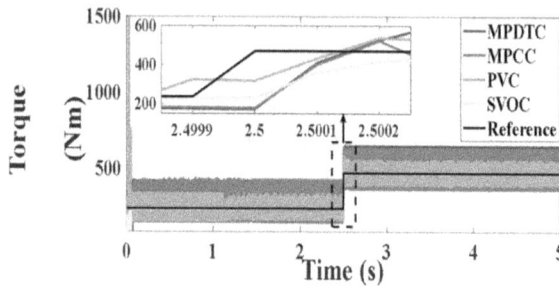

FIGURE 4.226 Comparison between torques of four techniques (Nm).

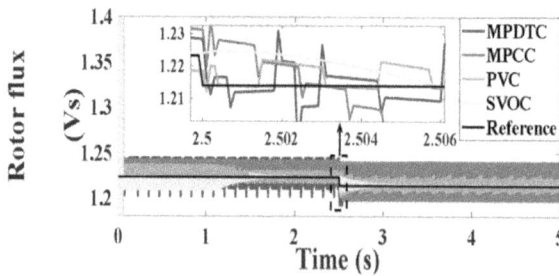

FIGURE 4.227 Comparison between rotor fluxes of four techniques (Vs).

TABLE 4.4

Response Time Taken by the Actual Values to Reach to their Reference Values in Second Case

Technique	Time Taken by the Active Power Profile (ms)	Time Taken by the Torque Profile (ms)	Taken by Rotor Flux Profile (ms)
SVOC	0.55	0.65	6.5
MPCC	0.16065	0.156	2.4
MPDTC	0.1604	0.151	0.14
PVC	0.14	0.135	0.057

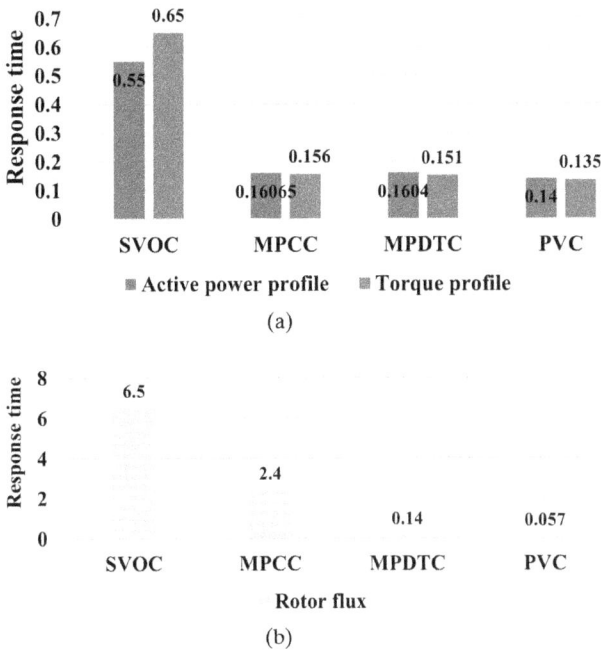

FIGURE 4.228 Histogram of response time taken by the actual values to reach to their reference values in second condition (s). (a) Profiles of active power and developed torque. (b) Rotor flux profile.

TABLE 4.5
Ripples Content of the Actual Values above their reference Values

Technique		SVOC	MPCC	MPDTC	PVC
Ripples of active power (Watt)	Case 1	740	7,040	7,890	2,980
	Case 2	670	7,580	8,140	3,070
Ripples of reactive power (Var)	Case 1	274	7,865	10,080	4,043
	Case 2	281	7,673	9,846	3,804
Ripples of developed torque (Nm)	Case 1	9.2	171.1	193.3	85
	Case 2	9.9	183.4	185	104.1
Ripples of rotor flux (Vs)	Case 1	0.012	0.023	0.024	0.016
	Case 2	0.013	0.02	0.022	0.014

(a)

(b)

(c)

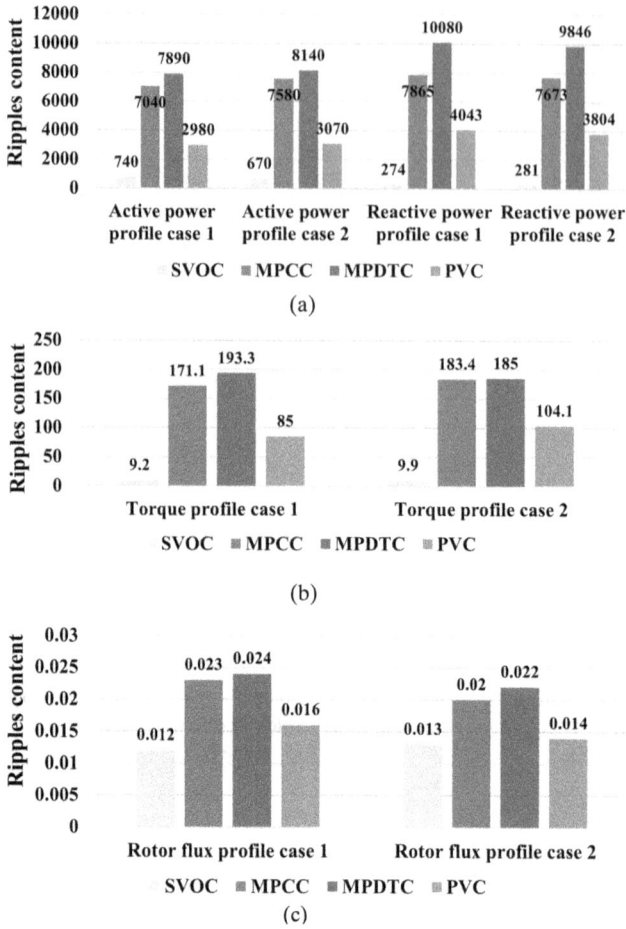

FIGURE 4.229 Histogram of ripples content of the actual values above their references. (a) Profiles of active power (Watt) and reactive power (Var). (b) Torque profile (Nm). (c) Rotor flux profile (Vs).

TABLE 4.6
Comparison in Terms of Performed Commutations by the Predictive Controllers

Technique	Variable Speed Operation	Fixed Speed Operation
MPCC	8,830	3,732
MPDTC	8,588	2,461
Proposed PVC	7,057	1,409

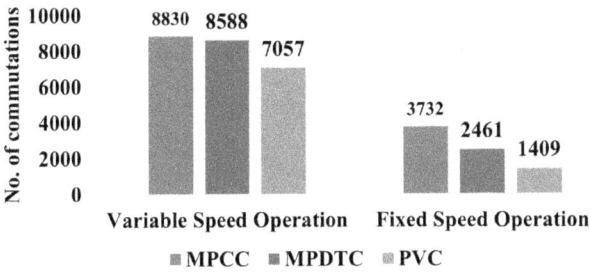

FIGURE 4.230 Histogram of number of performed commutations by the predictive controllers.

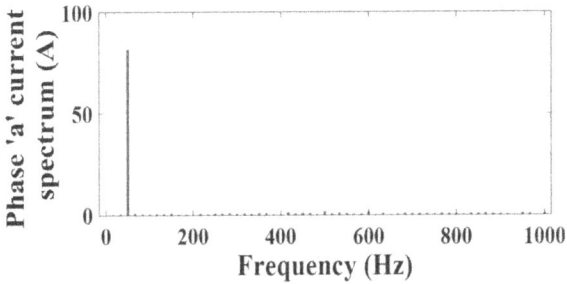

FIGURE 4.231 THD of Phase "a" of stator current with MPCC in first condition.

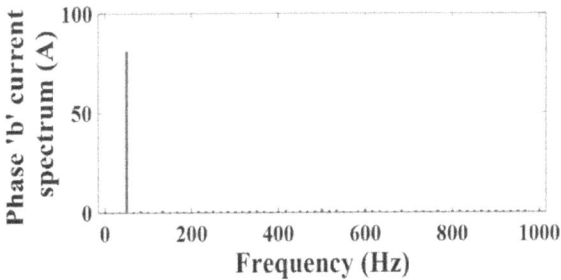

FIGURE 4.232 THD of Phase "b" of stator current with MPCC in first condition.

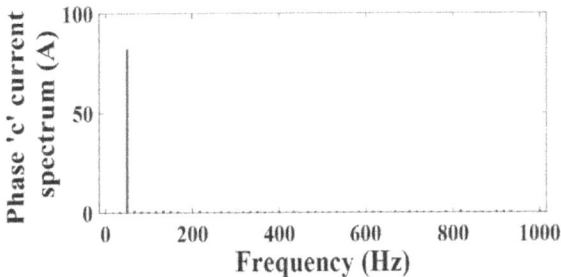

FIGURE 4.233 THD of Phase "c" of stator current with MPCC in first condition.

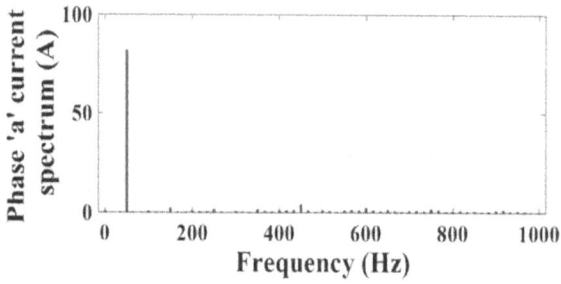

FIGURE 4.234   THD of Phase "a" of stator current with MPCC in second condition.

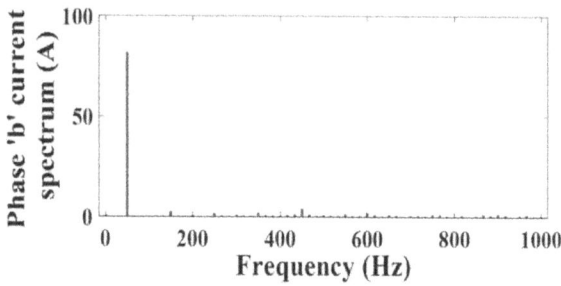

FIGURE 4.235   THD of Phase "b" of stator current with MPCC in second condition.

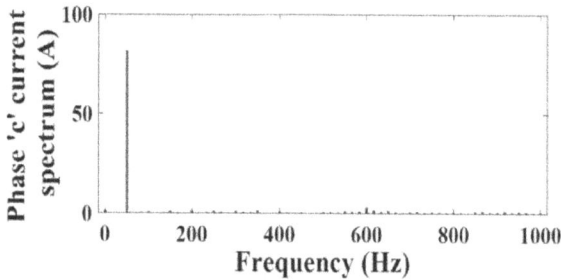

FIGURE 4.236   THD of Phase "c" of stator current with MPCC in second condition.

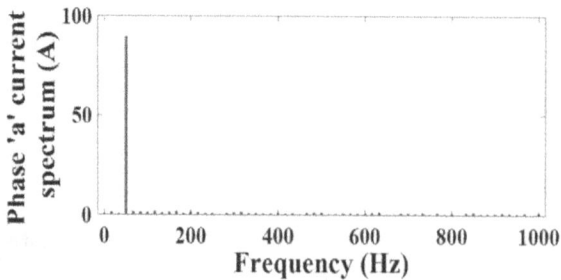

FIGURE 4.237   THD of Phase "a" of stator current with MPDTC in first condition.

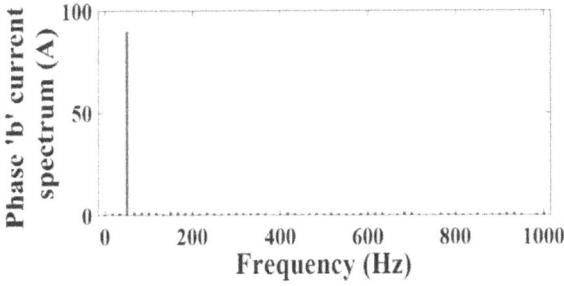

FIGURE 4.238   THD of Phase "b" of stator current with MPDTC in first condition.

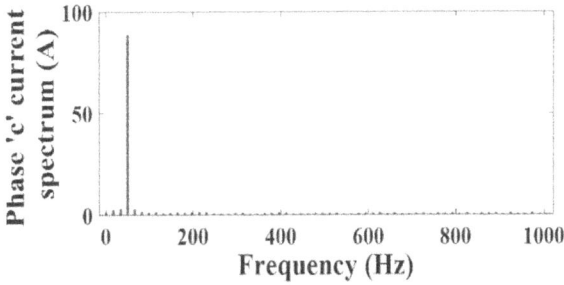

FIGURE 4.239   THD of Phase "c" of stator current with MPDTC in first condition.

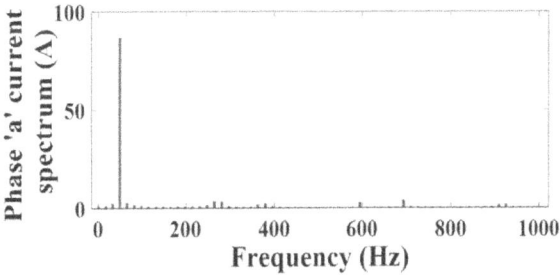

FIGURE 4.240   THD of Phase "a" of stator current with MPDTC in second condition.

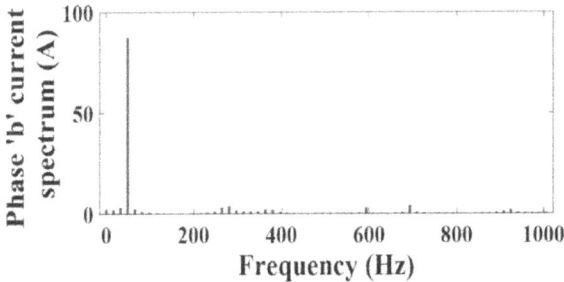

FIGURE 4.241   THD of Phase "b" of stator current with MPDTC in second condition.

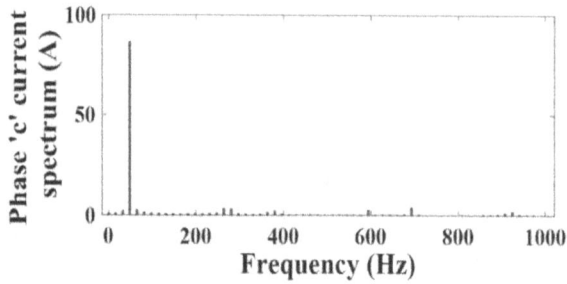

FIGURE 4.242   THD of Phase "c" of stator current with MPDTC in second condition.

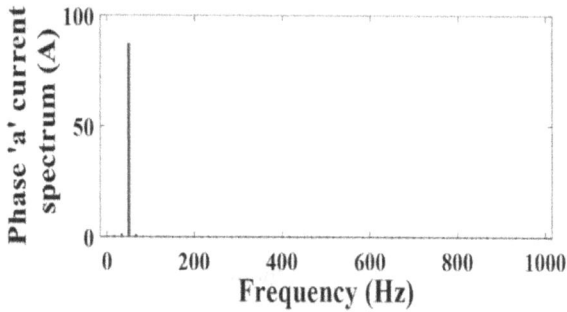

FIGURE 4.243   THD of Phase "a" of stator current with PVC in first condition.

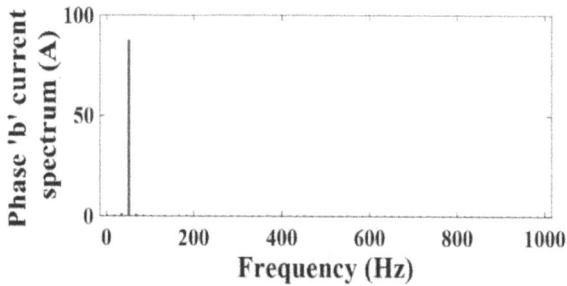

FIGURE 4.244   THD of Phase "b" of stator current with PVC in first condition.

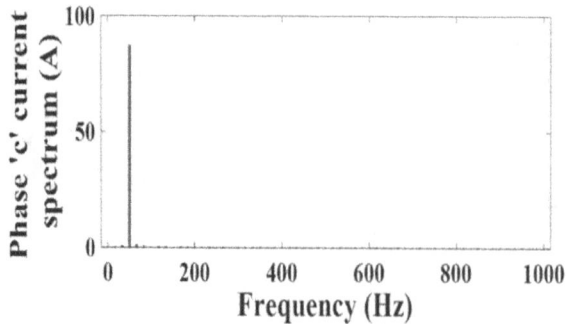

FIGURE 4.245   THD of Phase "c" of stator current with PVC in first condition.

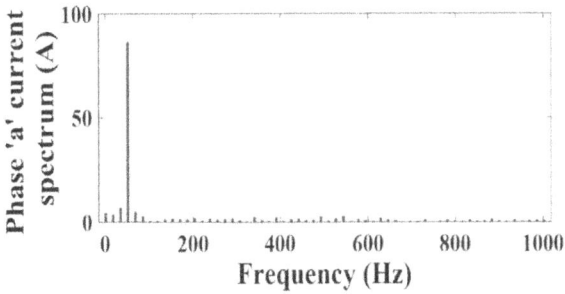

FIGURE 4.246 THD of Phase "a" of stator current with PVC in second condition.

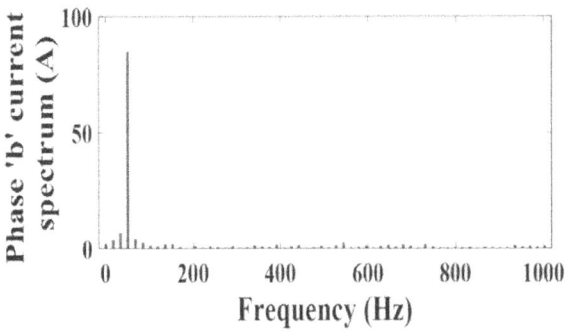

FIGURE 4.247 THD of Phase "b" of stator current with PVC in second condition.

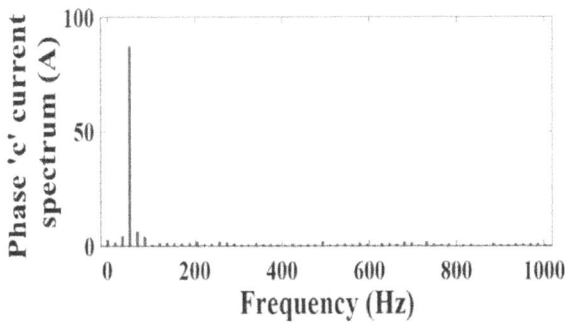

FIGURE 4.248 THD of Phase "c" of stator current with PVC in second condition.

TABLE 4.7

FFT Analysis for the Stator Current Components

	Technique			MPCC	MPDTC	PVC
Phase A	*Case 1*	Fundamental		81.6219 A	89.4056 A	87.2442 A
		THD		2.71%	3.09%	0.58%
	Case 2	Fundamental		81.8378 A	86.6527 A	86.2447 A
		THD		5.50%	6.02%	3.97%
Phase B	*Case 1*	Fundamental		81.0882 A	89.8613 A	87.3644 A
		THD		2.23%	3.25%	0.64%
	Case 2	Fundamental		81.5973 A	87.3355 A	84.6824 A
		THD		5.16%	5.97%	3.39%
Phase C	*Case 1*	Fundamental		82.0234 A	88.415 A	87.2582 A
		THD		2.03%	3.02%	0.31%
	Case 2	Fundamental		81.386 A	86.6246 A	87.0403 A
		THD		5.26%	6.19%	3.83%

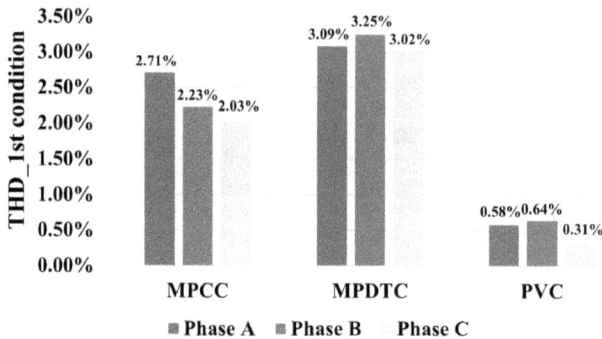

(a)

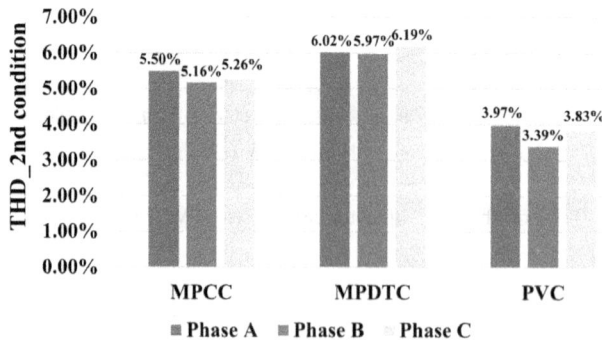

(b)

FIGURE 4.249 Histogram of THD analysis for the stator current components. (a) First condition. (b) Second condition.

5 Conclusions and Recommendations for Future Work

5.1 CONCLUSIONS

The following is a summary of the conclusions and contributions which were deduced from the book:

- The book introduced a modeling for the WECS including a wind turbine as a prime mover, a DFIG as a generation unit for electrical energy, and a three-phase induction motor as an isolated load.
- Enhancing and controlling the dynamic performance of a wind-driven DFIG during variable wind speeds using advanced control techniques.
- Presenting a detailed description for the presented control algorithms in order to visualize the base principle of each method, showing when it works properly and when it fails.
- Formulating an effective PVC approach which overcomes the shortages present in previous DFIG control schemes.
- Introducing a comprehensive performance analysis for the DFIG using the formulated PVC scheme and other control techniques under different operating conditions.
- The obtained results reveal and confirm the superiority of the formulated PVC over the other control schemes in terms of the fast dynamic response, simplicity, reduced ripples, THD, and computational burdens.
- Performing a fair and detailed comparison between different control topologies for the DFIG to outline the most effective procedure in terms of dynamic response, structure simplicity, ripples, THD, and computational burdens.
- The results of the comparison prove the effectiveness and validation of our proposed predictive controller; as it achieves the system simplicity, its dynamic response is faster than that of MPDTC and MPCC, it presents a lower content of ripples compared with MPDTC and MPCC. Moreover, it can minimize the computational burden and THD, remarkably. For these reasons, it can be said that the formulated controller is the most convenient to be used with the DFIG to achieve the best dynamic performance.
- The proposed PVC proves its enhanced robustness against parameter variation compared with recent control schemes that adopt an equivalent operating theory.

DOI: 10.1201/9781003440529-6

- The robustness of our designed controller against parameters variation arising from unreliability of its cost function on estimated variables, as the variables of the adopted cost function don't depend directly on estimated variables. The indirect relationship comes from using PI regulators instead of calculating the reference voltages directly from the machine model as in case of deadbeat principle. Also, we exploit the advantages of PI which make a buffering for the input signal which inherently holds and isolate the signal, so it doesn't directly transfer changes.
- The designed PVC scheme can be used with other machine configurations after considering the structure and operation theory of each type.

5.2 RECOMMENDATIONS FOR FUTURE WORK

In the near future, we will study the following topics, Insha'Allah:

- Studying the fault tolerance control of the wind-driven standalone DFIG.
- Extension the implementation of the proposed control algorithm to other generator types used with wind energy and also with tidal energy.
- Incorporating other generation units like solar cells or fuel cells with the wind generation system adopted in this book, in order to construct an effective hybrid generation system which can be used for various applications, particularly which require high power.
- Using of artificial intelligence like metaheuristic optimization algorithm and neural network in tuning the PI parameters.

Appendix A

TABLE A1
Data Specification of the DFIG and Wind Turbine

Parameter	Value
R_s	70 mΩ
R_r	87 mΩ
L_s	16.25 mH
L_r	16.3 mH
L_m	16 mH
Pole pairs (p)	3
Inertia $\left(J_g\right)$	0.3125 kg/m²
Stator rated voltage	380 V
Operating frequency	50 Hz
Rated power	1.5 MW
R	35.25 m
J_t	1,000 kg/m²
$C_{p,\max}$	0.51
λ_{opt}	8.1
G	90

TABLE A2
Parameters of the Load (Standalone Direct-Driven Operation)

Parameter	Value
Rated power	15 kW
Pole pairs $\left(p_{IM}\right)$	2
N_{rated}	1,400 rpm
T_{rated}	102 Nm

TABLE A3
Parameters of the Load (Standalone Wind-Driven Operation)

Parameter	Value
Rated power	400 kW
Pole pairs (p_{IM})	2
N_{rated}	1,400 rpm
T_{rated}	2,728 Nm

TABLE A4
Control System Parameters (Standalone Direct- & Wind-Driven Operation)

Parameter	Value
Sampling time	100 μs
DC link voltage (U_{dc})	570 V
K_p and K_i (Active power regulator)	0.0001 and −0.1
K_p and K_i (Stator voltage regulator)	0.01 and −20

TABLE A5
Control System Parameters (Grid-Connected Direct- & Wind-Driven Operation)

Parameter	Value
Sampling time	100 μs
DC link voltage (U_{dc})	210 V
K_p and K_i (d-axis rotor current regulator)	3 and 100
K_p and K_i (q-axis rotor current regulator)	3 and 100

References

[1] C. M. R. Charles, V. Vinod, and A. Jacob, "Field Oriented Control of DFIG Based Wind Energy System Using Battery Energy Storage System," *Procedia Technology*, vol. 24, pp. 1203–1210, 2016.

[2] J. R. Andrade and R. J. Bessa, "Improving Renewable Energy Forecasting With a Grid of Numerical Weather Predictions," *IEEE Transactions on Sustainable Energy*, vol. 8, no. 4, pp. 1571–1580, 2017.

[3] P. Siano, P. Chen, Z. Chen, and A. Piccolo, "Evaluating Maximum Wind Energy Exploitation in Active Distribution Networks," *IET Generation, Transmission & Distribution*, vol. 4, no. 5, pp. 598–608. Available: https://digital-library.theiet.org/content/journals/10.1049/iet-gtd.2009.0548

[4] M. A. Mossa, O. Gam, and N. Bianchi, "Performance Enhancement of a Hybrid Renewable Energy System Accompanied with Energy Storage Unit Using Effective Control System," *International Journal of Robotics and Control Systems*, vol. 2, no. 1, pp. 140-171. 2022. doi: 10.31763/ijrcs.v2i1.599

[5] M. A. Mossa, H. Echeikh, N. V. Quynh, and N. Bianchi, "Performance Dynamics Improvement of a Hybrid Wind/Fuel Cell/Battery System for Standalone Operation," *IET Renewable Power Generation*, vol. n/a, no. n/a, 2022. doi: 10.1049/rpg2.12603

[6] M. A. Mossa and S. Bolognani, "High Performance Direct Power Control for a Doubly Fed Induction Generator," In *IECON 2016 – 42nd Annual Conference of the IEEE Industrial Electronics Society*, Florence, Italy, pp. 1930–1935, 2016.

[7] A. A. Shaltout and M. A. Abdel-Halim, "Solid-State Control of a Wind-Driven Self-Excited Induction Generator," *Electric Machines & Power Systems*, vol. 23, no. 5, pp. 571–582, 1995.

[8] M. A. Mossa, O. Gam, and N. Bianchi, "Dynamic Performance Enhancement of a Renewable Energy System for Grid Connection and Stand-Alone Operation with Battery Storage," *Energies*, vol. 15, no. 3. doi: 10.3390/en15031002

[9] L. A. C. Lopes and R. G. Almeida, "Wind-Driven Self-Excited Induction Generator with Voltage and Frequency Regulated by a Reduced-Rating Voltage Source Inverter," *IEEE Transactions on Energy Conversion*, vol. 21, no. 2, pp. 297–304, 2006.

[10] M. M. Mahmoud, M. M. Aly, H. S. Salama, and A.-M. M. Abdel-Rahim, "Dynamic Evaluation of Optimization Techniques-Based Proportional-Integral Controller for Wind-Driven Permanent Magnet Synchronous Generator," *Wind Engineering*, vol. 45, no. 3, pp. 696–709, 2020.

[11] L. Dambrosio and B. Fortunato, "One-Step-Ahead Adaptive Control of a Wind-Driven, Synchronous Generator System," *Energy*, vol. 24, no. 1, pp. 9–20, 1999.

[12] J. Tian, D. Zhou, C. Su, F. Blaabjerg, and Z. Chen, "Maximum Energy Yield Oriented Turbine Control in PMSG-Based Wind Farm," *The Journal of Engineering*, vol. 2017, no. 13, pp. 2455–2460, 2017. doi: 10.1049/joe.2017.0770

[13] M. M. Mahmoud, B. S. Atia, A. Y. Abdelaziz, and N. A. N. Aldin, "Dynamic Performance Assessment of PMSG and DFIG-Based WECS with the Support of Manta Ray Foraging Optimizer Considering MPPT, Pitch Control, and FRT Capability Issues," *Processes*, vol. 10, no. 12. doi: 10.3390/pr10122723

[14] A. A. Hassan, A. M. El-Sawy, and O. M. Kamel, "Direct Torque Control of a Doubly fed Induction Generator Driven by a Variable Speed Wind Turbine," *Journal of Engineering Sciences*, vol. 41, no. 1, pp. 199–216, 2013.

[15] M. A. Mossa, O. Gam, N. Bianchi, and N. V. Quynh, "Enhanced Control and Power Management for a Renewable Energy-Based Water Pumping System," *IEEE Access*, vol. 10, pp. 36028–36056, 2022.

[16] O. M. Kamel, A. Y. Abdelaziz, and A. A. Zaki Diab, "Damping Oscillation Techniques for Wind Farm DFIG Integrated into Inter-Connected Power System," *Electric Power Components and Systems*, vol. 48, no. 14–15, pp. 1551–1570, 2020.

[17] S. Mahfoud, A. Derouich, N. El Ouanjli, M. El Mahfoud, and M. Taoussi, "A New Strategy-Based PID Controller Optimized by Genetic Algorithm for DTC of the Doubly Fed Induction Motor," *Systems*, vol. 9, no. 2, p. 37. doi: 10.3390/systems9020037

[18] M. A. Mossa and Y. S. Mohamed, "Novel Scheme for Improving the Performance of a Wind Driven Doubly Fed Induction Generator during Grid Fault," *Wind Engineering*, vol. 36, no. 3, pp. 305–334, 2012.

[19] A. Y. Abdelaziz, A. M. Ibrahim, A. M. Asim, A. H. A. Razek, and Y. G. Hegazy, "Investigation of Electrical Dynamics of DFIG-Based Wind Turbines During Severe Symmetrical Grid Voltage Dips," in *2012 International Conference on Engineering and Technology (ICET)*, Cairo, Egypt, pp. 1–6, 2012.

[20] M. R. A. Kashkooli, S. M. Madani, and T. A. Lipo, "Improved Direct Torque Control for a DFIG under Symmetrical Voltage Dip with Transient Flux Damping," *IEEE Transactions on Industrial Electronics*, vol. 67, no. 1, pp. 28–37, 2020.

[21] A. Gundavarapu, H. Misra, and A. K. Jain, "Direct Torque Control Scheme for DC Voltage Regulation of the Standalone DFIG-DC System," *IEEE Transactions on Industrial Electronics*, vol. 64, no. 5, pp. 3502–3512, 2017.

[22] R. M. Prasad and M. A. Mulla, "Mathematical Modeling and Position-Sensorless Algorithm for Stator-Side Field-Oriented Control of Rotor-Tied DFIG in Rotor Flux Reference Frame," *IEEE Transactions on Energy Conversion*, vol. 35, no. 2, pp. 631–639, 2020.

[23] M. A. Mossa, A. S. Al-Sumaiti, T. D. Do, and A. A. Z. Diab, "Cost-Effective Predictive Flux Control for a Sensorless Doubly Fed Induction Generator," *IEEE Access*, vol. 7, pp. 172606–172627, 2019.

[24] M. A. Mossa, H. Echeikh, A. A. Z. Diab, and N. V. Quynh, "Effective Direct Power Control for a Sensor-Less Doubly Fed Induction Generator with a Losses Minimization Criterion," *Electronics*, vol. 9, no. 8, p. 1269, 2020. doi: 10.3390/electronics9081269

[25] S. Mahfoud, A. Derouich, N. El Ouanjli, T. Mohammed, and A. Hanafi, "Field Oriented Control of Doubly Fed Induction Motor Using Speed Sliding Mode Controller," in *E3S Web Conference*, Agadir, Morocco, vol. 229, Article ID 01061, 2021. doi: 10.1051/e3sconf/202122901061

[26] F. Amrane, A. Chaiba, B. Francois, and B. Babes, "Experimental Design Of Stand-Alone Field Oriented Control for WECS in Variable Speed DFIG-Based on Hysteresis Current Controller," in *2017 15th International Conference on Electrical Machines, Drives and Power Systems (ELMA)*, Sofia, Bulgaria, pp. 304–308, 2017.

[27] S. El Daoudi, L. Lazrak, N. El Ouanjli, and M. Ait Lafkih, "Sensorless Fuzzy Direct Torque Control of Induction Motor with Sliding Mode Speed Controller," *Computers & Electrical Engineering*, vol. 96, p. 107490, 2021.

[28] C. Cheng and H. Nian, "Low-Complexity Model Predictive Stator Current Control of DFIG under Harmonic Grid Voltages," *IEEE Transactions on Energy Conversion*, vol. 32, no. 3, pp. 1072–1080, 2017.

[29] A. J. S. Filho, A. L. D. Oliveira, L. L. Rodrigues, E. C. M. Costa, and R. V. Jacomini, "A Robust Finite Control Set Applied to the DFIG Power Control," *IEEE Journal of Emerging and Selected Topics in Power Electronics*, vol. 6, no. 4, pp. 1692–1698, 2018.

[30] M. A. Mossa and S. Bolognani, "Robust Predictive Current Control for a Sensorless IM Drive Based on Torque Angle Regulation," in *2019 IEEE Conference on Power Electronics and Renewable Energy (CPERE)*, Aswan, Egypt, pp. 302–308, 2019.

[31] A. Bonfiglio *et al.*, "An MPC-Based Approach for Emergency Control Ensuring Transient Stability in Power Grids with Steam Plants," *IEEE Transactions on Industrial Electronics*, vol. 66, no. 7, pp. 5412–5422, 2019.

[32] T. Morstyn, B. Hredzak, R. P. Aguilera, and V. G. Agelidis, "Model Predictive Control for Distributed Microgrid Battery Energy Storage Systems," *IEEE Transactions on Control Systems Technology*, vol. 26, no. 3, pp. 1107–1114, 2018.

[33] M. Liu, K. W. Chan, J. Hu, W. Xu, and J. Rodriguez, "Model Predictive Direct Speed Control with Torque Oscillation Reduction for PMSM Drives," *IEEE Transactions on Industrial Informatics*, vol. 15, no. 9, pp. 4944–4956, 2019.

[34] S. G. Petkar and V. K. Thippiripati, "Enhanced Predictive Current Control of PMSM Drive with Virtual Voltage Space Vectors," *IEEE Journal of Emerging and Selected Topics in Industrial Electronics*, vol. 3, no. 3, pp. 834–844, 2022.

[35] Y. Zhang, J. Zhu, and J. Hu, "Model Predictive Direct Torque Control for Grid Synchronization of Doubly Fed Induction Generator," in *2011 IEEE International Electric Machines & Drives Conference (IEMDC)*, Niagara Falls, ON, Canada, pp. 765–770, 2011.

[36] M. M. Vayeghan and S. A. Davari, "Torque Ripple Reduction of DFIG by a New and Robust Predictive Torque Control Method," *IET Renewable Power Generation*, vol. 11, no. 11, pp. 1345–1352, 2017. doi: 10.1049/iet-rpg.2016.0695

[37] X. Wei, M. Cheng, J. Zhu, H. Yang, and R. Luo, "Finite-Set Model Predictive Power Control of Brushless Doubly Fed Twin Stator Induction Generator," *IEEE Transactions on Power Electronics*, vol. 34, no. 3, pp. 2300–2311, 2019.

[38] Y. Zhang, J. Jiao, D. Xu, D. Jiang, Z. Wang, and C. Tong, "Model Predictive Direct Power Control of Doubly Fed Induction Generators under Balanced and Unbalanced Network Conditions," *IEEE Transactions on Industry Applications*, vol. 56, no. 1, pp. 771–786, 2020.

[39] S. A. Davari, V. Nekoukar, C. Garcia, and J. Rodriguez, "Online Weighting Factor Optimization by Simplified Simulated Annealing for Finite Set Predictive Control," *IEEE Transactions on Industrial Informatics*, vol. 17, no. 1, pp. 31–40, 2021.

[40] M. E. Zarei, C. V. Nicolás, and J. R. Arribas, "Improved Predictive Direct Power Control of Doubly Fed Induction Generator During Unbalanced Grid Voltage Based on Four Vectors," *IEEE Journal of Emerging and Selected Topics in Power Electronics*, vol. 5, no. 2, pp. 695–707, 2017.

[41] A. M. Bozorgi, M. Farasat, and S. Jafarishiadeh, "Model Predictive Current Control of Surface-Mounted Permanent Magnet Synchronous Motor with Low Torque and Current Ripple," *IET Power Electronics*, vol. 10, no. 10, pp. 1120–1128, 2017. doi: 10.1049/iet-pel.2016.0850

[42] M. A. Mossa, N. V. Quynh, H. Echeikh, and T. D. Do, "Deadbeat-Based Model Predictive Voltage Control for a Sensorless Five-Phase Induction Motor Drive," *Mathematical Problems in Engineering*, vol. 2020, p. 4164526, 2020.

[43] J. Holtz, "A Predictive Controller for the Stator Current Vector of AC Machines Fed from a Switched Voltage Source," in *Proceedings of the IEE of Japan IPEC-Tokyo'83*, IPEC-Tokyo, pp. 1665–1675, 1983.

[44] D. Das, S. Das, and N. Mohammad, "Modified Model Predictive Current Control of Doubly Fed Induction Generator in Wind Energy Conversion System," in *2019 IEEE International Conference on Power, Electrical, and Electronics and Industrial Applications*, Dhaka, Bangladesh, pp. 48–52, 2019.

[45] R. Btm Consult ApS, "International Wind Energy Development World Market Update 2005 Forecast 2006–2010," Denmark87-987788-8-9, 2006, Available: https://inis.iaea.org/search/search.aspx?orig_q=RN:38045808.

[46] B. H. Chowdhury and S. Chellapilla, "Double-Fed Induction Generator Control for Variable Speed Wind Power Generation," *Electric Power Systems Research*, vol. 76, no. 9, pp. 786–800, 2006.

[47] A. Miller, E. Muljadi, and D. S. Zinger, "A Variable Speed Wind Turbine Power Control," *IEEE Transactions on Energy Conversion*, vol. 12, no. 2, pp. 181–186, 1997.

[48] J. Richalet, A. Rault, J. L. Testud, and J. Papon, "Model Predictive Heuristic Control: Applications to Industrial Processes," *Automatica*, vol. 14, no. 5, pp. 413–428, 1978.

[49] L. M. Fernández, F. Jurado, and J. R. Saenz, "Aggregated Dynamic Model for Wind Farms with Doubly Fed Induction Generator Wind Turbines," *Renewable Energy*, vol. 33, no. 1, pp. 129–140, 2008.

[50] M. Morari and J. H. Lee, "Model predictive control: past, present and future," *Computers & Chemical Engineering*, vol. 23, no. 4, pp. 667–682, 1999/05/01/ 1999.

[51] E. Muljadi and C. P. Butterfield, "Pitch-Controlled Variable-Speed Wind Turbine Generation," *IEEE Transactions on Industry Applications*, vol. 37, no. 1, pp. 240–246, 2001.

[52] H. M. Jabr and N. C. Kar, "Fuzzy Gain Tuner for Vector Control of Doubly-Fed Wind Driven Induction Generator," in *2006 Canadian Conference on Electrical and Computer Engineering*, Ottawa, ON, Canada, pp. 2266–2269, 2006.

[53] R. Spée, S. Bhowmik, and J. H. R. Enslin, "Novel Control Strategies for Variable-Speed Doubly Fed Wind Power Generation Systems," *Renewable Energy*, vol. 6, no. 8, pp. 907–915, 1995.

[54] J. C. Dai, Y. P. Hu, D. S. Liu, and J. Wei, "Modelling and Analysis of Direct-Driven Permanent Magnet Synchronous Generator Wind Turbine Based on Wind-Rotor Neural Network Model," *Proceedings of the Institution of Mechanical Engineers, Part A: Journal of Power and Energy*, vol. 226, no. 1, pp. 62–72, 2012.

[55] R. Pena, J. C. Clare, and G. M. Asher, "Doubly Fed Induction Generator Using Back-to-Back PWM Converters and Its Application to Variable-Speed Wind-Energy Generation," *IEE Proceedings – Electric Power Applications*, vol. 143, no. 3, pp. 231–241, 1996. Available: https://digital-library.theiet.org/content/journals/10.1049/ip-epa_19960288

[56] A. Tapia, G. Tapia, J. X. Ostolaza, and J. R. Saenz, "Modeling and Control of a Wind Turbine Driven Doubly Fed Induction Generator," *IEEE Transactions on Energy Conversion*, vol. 18, no. 2, pp. 194–204, 2003.

[57] A. M. A. Haidar, K. M. Muttaqi, and M. T. Hagh, "A Coordinated Control Approach for DC link and Rotor Crowbars to Improve Fault Ride-through of DFIG-Based Wind Turbine," *IEEE Transactions on Industry Applications*, vol. 53, no. 4, pp. 4073–4086, 2017.

[58] L. Xu, "Coordinated Control of DFIG's Rotor and Grid Side Converters During Network Unbalance," *IEEE Transactions on Power Electronics*, vol. 23, no. 3, pp. 1041–1049, 2008.

[59] R. M. Prasad and M. A. Mulla, "A Novel Position-Sensorless Algorithm for Field-Oriented Control of DFIG With Reduced Current Sensors," *IEEE Transactions on Sustainable Energy*, vol. 10, no. 3, pp. 1098–1108, 2019.

[60] B. B. R. Anita and B. Babypriya, "Modelling, Simulation and Analysis of Doubly Fed Induction Generator for Wind Turbines," *Journal of Electrical Engineering*, vol. 60, no. 2, pp. 79–85, 2009.

[61] D. W. Clarke, C. Mohtadi, and P. S. Tuffs, "Generalized Predictive Control – Part I. The Basic Algorithm," *Automatica*, vol. 23, no. 2, pp. 137–148, 1987.

[62] D. W. Clarke, C. Mohtadi, and P. S. Tuffs, "Generalized Predictive Control – Part II Extensions and Interpretations," *Automatica*, vol. 23, no. 2, pp. 149–160, 1987.

[63] M. A. Mossa and S. Bolognani, "Implicit Predictive Flux Control for High-Performance Induction Motor Drives," *Electrical Engineering*, vol. 103, no. 1, pp. 373–395, 2021.

[64] M. A. Mossa, M. K. Abdelhamid, A. A. Hassan, and N. Bianchi, "Improving the Dynamic Performance of a Variable Speed DFIG for Energy Conversion Purposes Using an Effective Control System," *Processes*, vol. 10, no. 3, p. 456, 2022.

[65] L. L. Rodrigues, O. A. C. Vilcanqui, A. L. L. F. Murari, and A. J. S. Filho, "Predictive Power Control for DFIG: A FARE-Based Weighting Matrices Approach," *IEEE Journal of Emerging and Selected Topics in Power Electronics*, vol. 7, no. 2, pp. 967–975, 2019.

[66] P. Kou, D. Liang, J. Li, L. Gao, and Q. Ze, "Finite-Control-Set Model Predictive Control for DFIG Wind Turbines," *IEEE Transactions on Automation Science and Engineering*, vol. 15, no. 3, pp. 1004–1013, 2018.

[67] A. M. Almaktoof, A. K. Raji, and M. T. E. Kahn, "Finite-Set Model Predictive Control and DC-Link Capacitor Voltages Balancing for Three-Level NPC Inverters," in *2014 16th International Power Electronics and Motion Control Conference and Exposition*, Antalya, Turkey, pp. 224–229, 2014.

[68] L. A. G. Gomez, L. F. N. Lourenço, A. P. Grilo, M. B. C. Salles, L. Meegahapola, and A. J. S. Filho, "Primary Frequency Response of Microgrid Using Doubly Fed Induction Generator With Finite Control Set Model Predictive Control Plus Droop Control and Storage System," *IEEE Access*, vol. 8, pp. 189298–189312, 2020.

[69] Z. Ortatepe and A. Karaarslan, "Error Minimization Based on Multi-Objective Finite Control Set Model Predictive Control for Matrix Converter in DFIG," *International Journal of Electrical Power & Energy Systems*, vol. 126, p. 106575, 2021.

[70] Y. A. Ali, M. Ouassaid, and E. Muljadi, "Reduced Switching Frequency Finite Control Set Model Predictive Control (FCS-MPC) for DFIG," in *2021 Innovations in Power and Advanced Computing Technologies (i-PACT)*, Kuala Lumpur, Malaysia, pp. 1–7, 2021.

[71] A. A. Ahmed, A. Bakeer, H. H. Alhelou, P. Siano, and M. A. Mossa, "A New Modulated Finite Control Set-Model Predictive Control of Quasi-Z-Source Inverter for PMSM Drives," *Electronics*, vol. 10, no. 22. doi: 10.3390/electronics10222814

[72] M. K. Abdelhamid, M. A. Mossa, and A. A. Hassan, "Enhancing the Dynamic Performance of a Standalone DFIG Under Variable Speed Operation Using an Effective Control Technique," *Journal of Advanced Engineering Trends (JAET)*, vol. 1, 2022.

[73] A. Zanelli, J. Kullick, H. M. Eldeeb, G. Frison, C. M. Hackl, and M. Diehl, "Continuous Control Set Nonlinear Model Predictive Control of Reluctance Synchronous Machines," *IEEE Transactions on Control Systems Technology*, vol. 30, no. 1, pp. 130–141, 2022.

[74] S. Vazquez, J. I. Leon, L. G. Franquelo, J. Rodriguez, H. A. Young, A. Marquez, and P. Zanchetta, "Model Predictive Control: A Review of Its Applications in Power Electronics," *IEEE Industrial Electronics Magazine*, vol. 8, no. 1, pp. 16–31, 2014.

[75] B. Hopfensperger, D. J. Atkinson, and R. A. Lakin, "Stator flux oriented control of a cascaded doubly-fed induction machine," *IEE Proceedings – Electric Power Applications*, vol. 146, no. 6, pp. 597–605. Available: https://digital-library.theiet.org/content/journals/10.1049/ip-epa_19990590

[76] M. Debbou, J. Gillet, T. Achour, and M. Pietrzak-David, "Control System for Doubly Fed Induction Machine in Electrical Naval Propulsion," in *2013 15th European Conference on Power Electronics and Applications (EPE)*, Lille, France, pp. 1–10, 2013.

[77] K. Spiteri, C. S. Staines, and M. Apap, "Control of Doubly Fed Induction Machine Using a Matrix Converter," in *MELECON 2010 – 2010 15th IEEE Mediterranean Electrotechnical Conference*, Valletta, Malta, pp. 1297–1302, 2010.

[78] P. Han, M. Cheng, and Z. Chen, "Single-Electrical-Port Control of Cascaded Doubly-Fed Induction Machine for EV/HEV Applications," *IEEE Transactions on Power Electronics*, vol. 32, no. 9, pp. 7233–7243, 2017.

[79] S. Peresada, A. Tilli, and A. Tonielli, "Indirect stator flux-oriented output feedback control of a doubly fed induction machine," *IEEE Transactions on Control Systems Technology*, vol. 11, no. 6, pp. 875–888, 2003.

[80] Q. Huang, X. Zou, D. Zhu, and Y. Kang, "Scaled Current Tracking Control for Doubly Fed Induction Generator to Ride-through Serious Grid Faults," *IEEE Transactions on Power Electronics*, vol. 31, no. 3, pp. 2150–2165, 2016.

[81] M. A. Mossa and S. Bolognani, "Predictive Power Control for a Linearized Doubly Fed Induction Generator Model," in *2019 21st International Middle East Power Systems Conference (MEPCON)*, Cairo, Egypt, pp. 250–257, 2019.

[82] M. A. Mossa, T. D. Do, A. S. Al-Sumaiti, N. V. Quynh, and A. A. Z. Diab, "Effective Model Predictive Voltage Control for a Sensorless Doubly Fed Induction Generator," *IEEE Canadian Journal of Electrical and Computer Engineering*, vol. 44, no. 1, pp. 50–64, 2021.

[83] M. K. Abdelhamid, M. A. Mossa, and A. A. Hassan, "Optimizing the Dynamic Performance of a Wind Driven Standalone DFIG Using an Advanced Control Algorithm," *Journal of Robotics and Control (JRC)*, vol. 3, no. 5, pp. 633–645, 2022.

[84] M. A. Mossa, H. Echeikh, and A. Iqbal, "Enhanced Control Technique for a Sensor-Less Wind Driven Doubly Fed Induction Generator for Energy Conversion Purpose," *Energy Reports*, vol. 7, pp. 5815–5833, 2021.

[85] J. Da Silva, R. De Oliveira, S. Silva, B. Rabelo, and W. Hofmann, "A Discussion about a Start-Up Procedure of a Doubly-Fed Induction Generator System," in *Nordic Workshop on Power and Industrial Electronics (NORPIE/2008)*, June 9–11, 2008, Espoo, Finland, Helsinki University of Technology, 2008.

[86] Y. Zhang and X. Donglin, "Direct Power Control of Doubly Fed Induction Generator Based on Extended Power Theory under Unbalanced Grid Condition," in *2017 IEEE 3rd International Future Energy Electronics Conference and ECCE Asia (IFEEC 2017 – ECCE Asia)*, Kaohsiung, Taiwan, pp. 992–996, 2017.

[87] C. M. R. Osorio, J. S. S. Chaves, A. L. L. F. Murari, and A. J. S. Filho, "Comparative Analysis of the Doubly Fed Induction Generator (DFIG) Under Balanced Voltage Sag Using a Deadbeat Controller," *IEEE Latin America Transactions*, vol. 15, no. 5, pp. 869–876, 2017.

[88] R. Franco, C. E. Capovilla, R. V. Jacomini, J. A. T. Altana, and A. J. S. Filho, "A Deadbeat Direct Power Control Applied to Doubly-Fed Induction Aerogenerator under Normal and Sag Voltages Conditions," in *IECON 2014 – 40th Annual Conference of the IEEE Industrial Electronics Society*, Dallas, TX, pp. 1906–1911, 2014.

Index

Note: **Bold** page numbers refer to tables and *italic* page numbers refer to figures.

For Product Safety Concerns and Information please contact our EU
representative GPSR@taylorandfrancis.com
Taylor & Francis Verlag GmbH, Kaufingerstraße 24, 80331 München, Germany

www.ingramcontent.com/pod-product-compliance
Lightning Source LLC
Chambersburg PA
CBHW070721220326
41598CB00024BA/3252

* 9 7 8 1 0 3 2 5 7 6 8 5 5 *